Evolutionary Studies

Series Editor

Naruya Saitou, National Institute of Genetics, Mishima, Japan

Everything is history, starting from the Big Bang or the origin of the universe to the present time. This historical nature of the universe is clear if we look at evolution of organisms. Evolution is one of most basic features of life which appeared on Earth more than 3.7 billion years ago. Considering the importance of evolution in biology, we are inaugurating this series. Any aspect of evolutionary studies on any kind of organism is a potential target of the series. Life started at the molecular level, thus molecular evolution is one important area in the series, but non-molecular studies are also within its scope, especially those studies on evolution of multicellular organisms. Evolutionary phenomena covered by the series include the origin of life, fossils in general, Earth–life interaction, evolution of prokaryotes and eukaryotes, viral and protist evolution, the emergence of multicellular organisms, phenotypic and genomic diversity of certain organism groups, and more. Theoretical studies on evolution are also covered within the spectrum of this new series.

Naoki Osada • Masahiko Kumagai •
Hitoshi Suzuki • Mitsuaki Endo
Editors

Phylogeographic History of Plants and Animals Coexisting with Humans in Asia

Springer

Editors
Naoki Osada
Faculty of Information Science and Technology
Hokkaido University
Sapporo, Hokkaido, Japan

Masahiko Kumagai
Research Center for Advanced Analysis
National Agriculture and Food Research Organization (NARO)
Tsukuba, Ibaraki, Japan

Hitoshi Suzuki
Graduate School of Environmental Earth Science
Hokkaido University
Sappro, Hokkaido, Japan

Mitsuaki Endo
School of Economics
Aoyama Gakuin Univesity
Shibuya-ku, Tokyo, Japan

ISSN 2509-484X ISSN 2509-4858 (electronic)
Evolutionary Studies
ISBN 978-981-97-6886-8 ISBN 978-981-97-6887-5 (eBook)
https://doi.org/10.1007/978-981-97-6887-5

This Springer imprint is published by the registered company Springer Nature Singapore Pte Ltd.
The registered company address is: 152 Beach Road, #21-01/04 Gateway East, Singapore 189721, Singapore

Preface

Genomics, archaeology, and linguistics. One of the shared key concepts of these three different disciplines is history. These disciplines, each distinct in its approach and methodology, emphasize the dynamic nature of our existence, highlighting that our genomes, cultures, and languages do not remain static but evolve over time. They become potent tools in revealing human history, especially when their analyses are coherently integrated. This synergistic approach allows for a more comprehensive understanding of how humans have adapted, migrated, and evolved through millennia, offering insights into the complex interplay between biology, culture, and communication. This book is to provide a comprehensive understanding of the natural history of Japan, using cutting-edge genomic research to reveal insights into the past.

The anticipated readership for this book encompasses both academic scholars and general enthusiasts with a keen interest in Japanese history, culture, and natural history. Additionally, professionals in fields such as biology, genetics, ecology, and linguistics, along with graduate students specializing in these disciplines, are among the intended audience. This book places significant emphasis on genomics, a field that has experienced remarkable advancements due to recent developments in genome sequencing techniques. Specifically, the analysis of ancient DNA has significantly enhanced our comprehension of the recent history of humans and other species. Nevertheless, research on ancient DNA from organisms within the Japanese archipelago has lagged slightly behind studies conducted in other regions, such as Europe and mainland China. In the near future, readers can anticipate a plethora of intriguing discoveries stemming from the utilization of ancient DNA data in the Japanese archipelago.

This book commences with an introductory chapter authored by four editors, offering an overview of its contents. Readers are encouraged to peruse this chapter initially and subsequently select any other chapter for further exploration. Within this introductory chapter, a table is provided summarizing the estimated migration origins and arrival times of species discussed throughout this book. This table serves as a valuable reference regarding the historical interactions between humans and the fauna and flora of the Japanese archipelago. It is important to note that the opinions expressed within this book are the sole responsibility of the editors, excluding the authors of individual chapters. With the exception of the final chapter, which presents a summary of linguistic correlations across the preceding chapters, each chapter stands independently. Readers have the flexibility to commence their reading with any chapter of their choosing, whether it pertains to rodents, bears, dogs, wolves, chickens, horses, taro, millet, rice, or wheat.

Many of the findings shared in this book stem from the Yaponesia Genome Project, which was supported by the Ministry of Education, Culture, Sports, Science, and Technology in Japan (MEXT) from 2018 to 2023. This project was led by Professor Naruya Saitou, an evolutionary biologist and anthropologist, and aimed to uncover the history of human settlement in Yaponesia—a term referring to the Japanese Archipelago that was inspired by Japanese novelist Toshio Shimao (1917–1986). This project brought together over 70 researchers from diverse fields, such as genomics, archaeology, and linguistics, across eight research groups. Many authors contributing to this book were part of the A3 research group, which focuses on the history of animals and plants in Yaponesia. We extend our gratitude to all project members for their contributions. We also invited researchers in these topics outside of the research projects, which make the contents of this book more comprehensive and insightful. We appreciate the efforts of all contributors to this book.

Sapporo, Hokkaido, Japan Naoki Osada

Contents

Chapter 1
Overview of the Book

Naoki Osada, Hitoshi Suzuki, Masahiko Kumagai, and Mitsuaki Endo

Keywords Animal · Plant · Domestication · Commensalism · Genome

1.1 Introduction

The main topic of this book is the evolution of animals and plants that are closely associated with human activities in East Asia, particularly focusing on the Japanese Archipelago (Yaponesia). During the Neolithic period, East Asians increasingly relied on plant-based diets, particularly millet and rice. This marked the beginning of agriculture, and East Asia became one of the world's major agricultural hubs. Following the initial arrival of modern humans in the Japanese Archipelago around 40,000 years ago, inhabitants extensively used various animal and plant resources. Although some species were native, many others were introduced to the archipelago by humans. Notably, the migration of the Yayoi people was influenced by

N. Osada (✉)
Faculty of Information Science and Technology, Hokkaido University, Sapporo, Hokkaido, Japan
e-mail: nosada@ist.hokudai.ac.jp

H. Suzuki
Graduate School of Environmental Science, Hokkaido University, Sapporo, Hokkaido, Japan
e-mail: htsuzuki@ees.hokudai.ac.jp

M. Kumagai
Research Center for Advanced Analysis, National Agriculture and Food Research Organization (NARO), Tsukuba, Ibaraki, Japan
e-mail: kumagai.m@naro.affrc.go.jp

M. Endo
School of Economics, Aoyama Gakuin University, Shibuya-ku, Tokyo, Japan
e-mail: t10976@aoyamagakuin.jp

N. Osada et al. (eds.), *Phylogeographic History of Plants and Animals Coexisting with Humans in Asia*, Evolutionary Studies,
https://doi.org/10.1007/978-981-97-6887-5_1

agricultural advancements on the continent, and they introduced the wet-field rice cultivation system, which profoundly altered the animal and plant landscape of the Japanese Archipelago. However, not all species were intentionally introduced; for example, rodents such as house mice and black rats, which live in close association with humans, quickly exploited new niches created by human activities and spread throughout the Japanese Archipelago. This influx of new species significantly transformed the interactions among humans, animals, and plants in the region.

This book illustrates the history of a diverse array of animal and plant species that have closely interacted with humans in East Asia. The information presented in this book draws mainly on genomic research, showcasing how advancements in genome sequencing technology have deepened our understanding of these species' histories. The analysis of ancient genomic data, particularly of animals, has been instrumental in tracing the formation of modern populations through their interactions with humans. Additionally, this book explores linguistic perspectives in its concluding chapter, offering readers an integrated view of the evolution of animals and plants in East Asia, enriched by a broad spectrum of evidence.

This book covers a wide range of animal and plant species associated with human activities. Animal species are discussed in Chaps. 2–6, whereas plant species are covered in Chaps. 7–10. Chapter 11 concludes the book with a summary of the geolinguistic patterns of terms for the animals and plants mentioned throughout the text. Table 1.1 provides a summary of the estimated arrival times and origins of each species discussed. Except for bears and Japanese barnyard millet, all featured animals and plants arrived in the Japanese Archipelago after the first human settlements. The following information provides a brief overview of the contents in each chapter.

1.2 Summary of Chaps. 2–6

Chapters 2–6 explore various animals and their distinct relationships with humans. Chapter 2, by Osada and Suzuki, investigated the evolutionary history of mice and rats. These rodents do not offer direct benefits to humans but have expanded their habitats alongside human migrations and trade, making them commensal species. Their origins and migration paths to Yaponesia were analyzed through mitochondrial and nuclear genome data.

Chapter 3, authored by Masuda, traced the genetic history of the brown bear, the largest terrestrial mammal in the Northern Hemisphere, and discussed its cultural significance in Northeast Asia. Particularly in Yaponesia, brown bears hold ritual importance among the indigenous Ainu people of Hokkaido Island. Genetic research revealed unique features and a complex migratory history of Hokkaido's brown bears to Yaponesia.

In Chapter 4, Terai addressed the history of the Japanese wolf, which has been extinct in Japan for over 120 years, and the origins of domestic dogs. This chapter built on recent findings, including ancient genome sequencing by Terai's team. The

Table 1.1 List of species discussed in this book and their approximate arrival times to the Japanese Archipelago

Common name	Linnaean name	Initial arrival time to Japan (years ago)	Homeland[a]	Chapter
House mouse	*Mus musculus*	4000–3000	South Asia→East Asia	2
Brown rat	*Rattus norvegicus*	?	East Asia or Southeast Asia	2
Black rat	*Rattus rattus/tanezumi*	2000–1000?	Southeast Asia	2
Brown bear	*Ursus arctos*	After 200,000	Inner Eurasian Continent	3
Dog	*Canis lupus familiaris*	11,500–9600	East Asia	4
Horse	*Equus caballus*	1500–1400	Mongolia	5
Chicken	*Gallus gallus*	2900–1700 (Yayoi period)	Southeast Asia→East Asia	6
Taro	*Colocasia esculenta*	After 40,000?	South Asia	7
Japanese barnyard millet	*Echinochloa esculenta*	–	Japan?	8
Proso millet	*Panicum miliaceum*	~2900	East Asia	8
Foxtail millet	*Setaria italica*	~2900	East Asia	8
Japonica rice	*Oryza sativa japonica*	~2900	East Asia	9
Common wheat	*Triticum aestivum*	~2000?	Middle East→East Asia	10

The beginning of the Yayoi period was assumed to be 2900 years ago
[a] Range of wild populations, migration origin or migration route of Japanese population

debate on dog domestication origins was enriched by evidence suggesting a close relationship between the Japanese wolf and domestic dogs, which provided valuable insights into the domestication process.

In Chapter 5, Tozaki discussed the genetic diversity and history of Japanese native horses through his latest research using genome-wide variant data. Horses have been integral to Japanese culture, serving roles in warfare, agriculture, and rituals. The presence of numerous horse statues with saddles discovered in mounds from the Kofun period (third to sixth centuries) underscores their cultural significance during these times. Genetic analyses indicated that native Japanese horses were derived from Mongolian horse populations and spread throughout Yaponesia following their initial introduction.

Chapter 6, by Yonezawa, delved into the evolutionary origins and history of chickens. Originating from Southeast Asia, chickens were introduced to Yaponesia for diverse purposes. By analyzing mitochondrial genomes, Yonezawa outlined a three-phase model for their introduction and revealed the intricate migration history of Japanese native chickens.

1.3 Summary of Chaps. 7–10

Chapters 7–10 deal with rice, millets, wheat, and taro, which were introduced in Yaponesia and are closely linked to people's livelihoods. Tracing the geographical or genetic origins of cultivated species is not straightforward. This is because of the drastic nature of the movement dynamics of cultivated species and the influence of ongoing genetic exchange with ancestral wild species or related species. Therefore, research efforts require a multidisciplinary approach that incorporates biogeographical analysis, archaeological evidence, ancient climate, and DNA analysis of ancient samples. Such attempts are currently underway to gain a comprehensive understanding of the history of these species.

In Chapter 7, Motohashi et al. discussed the origin and dissemination routes of taro, also known as "satoimo" in Japan. Taro is an ancient crop that predates the introduction of cultivated crops. The authors analyzed chloroplast DNA sequences and genome-wide SNP data from samples collected across various regions worldwide to elucidate the classification of varieties and genomic diversity. Furthermore, they provided a detailed discussion of the origins of Japanese lineages.

Chapter 8 by Satomura provided an overview of previous research on three millets: proso millet, foxtail millet, and barnyard millet, focusing on their relationships with wild species, origins, and historical spread across the Eurasian continent. This chapter reviews the genomic characteristics, distributions of cultivated and wild millets, and archaeological evidence, including ceramic impressions and changes in grain size, and summarizes the hypothesis of the dissemination of millets from Eurasia to Japan.

In Chapter 9, Kumagai and Sakai reviewed the debate surrounding the geographical origins of cultivated rice, one of the most extensively analyzed cereal species in terms of genomic data, as well as the complex relationship between japonica and indica. This chapter introduces insights into the geographical dispersal processes of population subgroups in Asia from recent population genomic analyses, as well as the genomic diversity of indigenous rice varieties in Japan.

In Chapter 10, Okada and Shimizu explained recent insights from genomic data analysis of the cultivation of wheat, the world's most widely cultivated species, including common wheat, einkorn, emmer, and timopheevi, focusing on complicated allopolyploidization events. Through population genetics studies, they discussed the origins of wheat cultivation and its dispersal processes both eastward and westward. This discussion was framed in the context of genomic science and archaeobotanical perspectives.

1.4 Summary of Chap. 11

Yaponesians Genome Project aimed to determine the origins of human groups in the Japanese archipelago through collaboration among geneticists, archaeologists, and linguists. The following sections examine the relevant points between the genetic and linguistic aspects of animals and crops, with a focus on Japanese and its adjacent languages.

Although no related words are found in the *Mus musculus musculus* area, some word forms, such as n- in southern China and Southeast Asia, may be related to *nezumi* in Japonic with certain probability. Japanese, Korean, and Chinese people use a mouse/rat euphemism, possibly due to crop damage concerns. In the vicinity of the mouse homeland, Iran, many languages distinguish between the terms rat and mouse. Such minute vocabulary differentiation is a common pattern observed in the center of production for livestock and crops.

The Ainu language in Hokkaido, one of the concentrating areas of bears, distinguishes between male, female, cub, and synecdoche terms for bears (Fukazawa 2022: 242–243). The words for "cub" in Iomante varied from Sakhalin to central Hokkaido and southern Hokkaido, mirroring the DNA distinction between the type in central Hokkaido and the other two types. Meanwhile, Chukotko–Kamchatkan languages use different terms for polar bears and brown bears, whereas Tibetan languages distinguish between brown bears and Asiatic black bears, where both species live. The word form *kuma* is common in Japanese, Korean, and Chinese.

Dogs are the same as mice/rats, and words across language families are scarce. In general, onomatopoeic words beginning with *k-* evolved independently within each language family.

In terms of the wolf, the Japanese word *ookami* translates to "great god." The *Chronicle of Japan* contains a story about two wolves who become gods. Such wolf worship is similar to that of northern horseback people, such as Mongolians.

The *mar*-like words for horse in East and Southeast Asia are typical examples of the so-called "Wanderwort," which is a sporadic word that spreads across language families. *Uma* "horse" is part of a group of words related to Chinese, Korean, and Mongolian. It already feels like an indigenous Japanese word because the borrowing date predates the main loan word layers in the eighth century AD.

In the case of chickens, the Japanese word *niwa-tori* is made up of two parts: "garden" and "bird." The sole *tori* also refers to chicken, a generic term for the newly domesticated species. Chinese and Kra–Dai share a common word form, *kei*. In most cases, scholars interpret minority languages as having adopted a major or dominant language. Because genetic evidence indicates that the chicken's native region is Southeast Asia, it is possible that the Chinese borrowed the Kra–Dai form. However, the major immigration of Kra–Dais to Southeast Asia dates back only to the Mongol quest period, which occurred approximately 700 years ago.

The terms for taro and yam are only found in Southeast and East Asia, as well as Bantu languages, indicating the objects' distribution. In many languages, yam is also known as taro. It is difficult to determine the commonality of word forms between these languages. In Japanese, the terms *sato* "village" and *yama* "mountain" are used before the generic term *imo* for root vegetable.

Broomcorn, foxtail, and barnyard millets have distinct names only in Chinese, Korean, and Japanese in East Asia, and the surrounding languages only use the generic term for millet. These three languages use a common term for barnyard millets, such as *hie* 稗 in Japanese. In northern Korea, this word refers to rice.

The Japanese word *mugi* for wheat may have been borrowed from Chinese. This word, like *uma*, is a loanword before eighth-century China. Because it is not cultivated in southern China, Kra–Dai and other Southeast and South Asian languages use loanwords that have only recently been adopted. Furthermore, because the climate is too cold to cultivate, the Ainu and Altaic type languages rely on loanwords from Chinese or Russian. This is consistent with the route for wheat propagation described by Shimizu and Okada in this volume.

The etymology of *ine* "rice plant" in Japanese is interpreted as "good food," whereas *kome* "rice grain" is derived from Austronesian *humay*. Millets migrated from Shandong to Taiwan. Contact between Austronesians and Japanese-speaking people may occur in Shandong province or further south.

In summary, bear, horse, barnyard millet, and wheat are common words in Japanese, Korean, and Chinese, despite being from different language families. Other matters and vocabulary vary depending on their homelands, the routes and duration of entry into Japan, as well as the situations and degrees affecting human daily life. Each language has different strategies for adapting to the introduction of new things. We will see their concrete aspects in Chap. 11.

Reference

Fukazawa, M (2022) 'Bear' in Ainu. In: Hiroyuki Suzuki, Mika Fukazawa, Akiko Yokoyama, and Mitsuaki Endo, eds. 2022. Linguistic Atlas of Asia and Africa, Vol. 1. Tokyo: Geolinguistic Society of Japan. https://doi.org/10.5281/zenodo.7118188

Chapter 2
Human Impacts on the Evolution of Rats and Mice in East Asia

Naoki Osada and Hitoshi Suzuki

Abstract Modern human activities have significantly reshaped Earth's ecosystems, notably through the domestication of animals and plants during the Neolithic era. This led to profound changes in the relationship between humans and other species, particularly in the development of agriculture across various regions. Rodents, including house mice (*Mus musculus*), brown rats (*Rattus norvegicus*), and black rats (*Rattus rattus* complex or *Rattus rattus/tanezumi*), exemplify species that have adapted to human-modified environments, achieving extensive global distribution. The paper focuses on the evolution of rats and mice in relation to human activities, particularly in East Asia and the Japanese Archipelago (Yaponesia). Advances in DNA sequencing technology have enabled more precise genetic histories for these species, supplementing traditional mitochondrial and microsatellite data. This has opened avenues for understanding their evolutionary history, especially as commensal animals, providing insights into pest control, laboratory research, and human cultural and biological history post-Neolithic era.

Keywords Genomics · House mouse · Black rat · Brown rat · Commensalism

Abbreviation

RrC Rattus rattus complex

N. Osada (✉)
Faculty of Information Science and Technology, Hokkaido University, Sapporo, Hokkaido, Japan
e-mail: nosada@ist.hokudai.ac.jp

H. Suzuki
Graduate School of Environmental Science, Hokkaido University, Sapporo, Hokkaido, Japan
e-mail: htsuzuki@ees.hokudai.ac.jp

N. Osada et al. (eds.), *Phylogeographic History of Plants and Animals Coexisting with Humans in Asia*, Evolutionary Studies,
https://doi.org/10.1007/978-981-97-6887-5_2

2.1 Introduction

Modern humans have significantly altered Earth's ecosystems, and one of the most transformative impacts is the domestication of animals and plants. While evidence suggests some management of wild species during the Paleolithic era (Douglass et al. 2021), the advent of agriculture and pastoralism in the Neolithic era transpired independently across various regions of the globe, fundamentally shifting human relationships with other organisms. Through the selective breeding of animals and plants for desirable traits, humans have not only manipulated but also enriched global biodiversity, engendering a multitude of species and varieties that had never previously existed in the natural world (Hulme-Beaman et al. 2016). However, many species thrive in anthropogenic environments and can often exhibit parasitic characteristics. While offering minimal advantages to humans, these commensal species have significantly expanded their natural habitats into human territories after ancestors of modern European and Asian populations migrated out of Africa (Bae et al. 2017).

Rodents, for example, pose substantial challenges to human health and economic stability by acting as pathogen carriers and agricultural pests. Specifically, three ubiquitous rodent groups of Asian origin, house mice (*Mus musculus*), brown rats (*Rattus norvegicus*), and black rats (*Rattus rattus/tanezumi*), have displayed remarkable adaptability to human-modified environments, achieving the most extensive global distribution of any other animals (Fig. 2.1). In Japanese, they are collectively referred to as "*ie* [ie] *nezumi*," where *nezumi* is a term encompassing rodents in general, with no distinct differentiation between mice and rats as is observed in some languages. Molecular dating studies have suggested that the genera *Mus* and

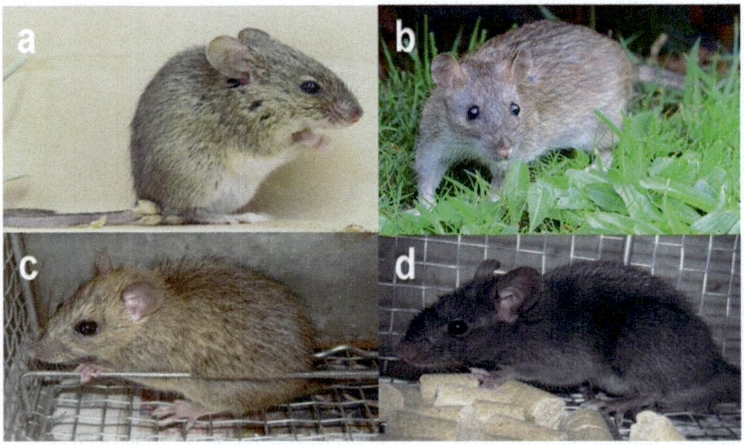

Fig. 2.1 Images of (**a**) *Mus musculus,* (**b**), *Rattus norvegicus* (**c**) *Rattus tanezumi* from Chichijima Island Ogasawara (**d**) melanistic *Rattus rattus* from Tokyo. (The panel **b** is licensed under the Creative Commons Attribution-Share Alike 4.0 International (https://creativecommons.org/licenses/by-sa/4.0/deed.en). The panels **c** and **d** are adopted from Kambe et al. (2011))

Rattus diverged approximately 12 Mya (Kumar et al. 2017). In East Asia, rodents are occasionally kept as pets, which led to the publication of books detailing breeding and care methods in Japan (Kuramoto 2011; Serikawa 2004; Zeng et al. 2018; Takada et al. 2013 #12498). Further, due to their ease of care and breeding, two of these species, *M. musculus* and *R. norvegicus*, are popularly bred as laboratory animals, contributing significantly to genetic and biomedical research and thus benefiting human health.

This paper offers a review of the current understanding of human impacts on the evolution of rats and mice, with a primary focus on East Asia, including the Japanese Archipelago (Yaponesia). The genetic diversity and geographic distributions of house mice, brown rats, and black rats are largely influenced by human activities. Although there is no evidence for the presence of *Mus* species in the Japanese Archipelago before the arrival of the ancestors of modern humans, whether the two *Rattus* species are endemic to the Japanese Archipelago remains controversial. This uncertainty persists despite existing fossil records, primarily because of difficulties and confusion in distinguishing rodent species (Kawamura 1988, 1989). Furthermore, their semi-fossorial lifestyle increases the likelihood of soil contamination, thus complicating the interpretation of fossil records. Given the challenges of tracing the presence of historical species, *R. norvegicus* is considered the most likely species present in the Japanese Archipelago during the Pleistocene (Kawamura 1989).

Recent advances in DNA sequencing technologies have paved the way for tracing the recent genetic history of species or populations using organismal DNA sequencing data, yielding significant insights into the history of many species, including mice and rats. Reference genome sequences for the three key species—house mice, brown rats, and black rats—have already been established (de Jong et al. 2022; Mouse Genome Sequencing 2002; Rane et al. 2020; Teng et al. 2016). However, their genetic diversity and pattern of differentiation at the genome-wide scale require further investigation. While studies have traditionally relied on mitochondrial genomes or microsatellite sequences, the advent of cost-effective whole-genome sequencing of nuclear genomes has expanded research capabilities. Mitochondrial genomes, which are haploids and lack recombination with other genomes, do not capture complex species histories, including admixture and hybridization events. Nuclear genome analysis, however, has effectively addressed these limitations. Furthermore, even with less-than-optimal coverage, whole-genome sequence data can be used to assemble complete mitochondrial genomes, which is particularly beneficial for studying animal genomes, such as those of rodents. Unraveling their detailed evolutionary history, especially their journey as commensal animals, offers valuable insights not only for pest control and laboratory utilization, but also for better understanding the biological and cultural history of humans during the post-Neolithic period (Jones et al. 2013).

2.2 Mouse

Overview

M. musculus (house mouse), the predominant commensal species within the genus *Mus*, has a vast global distribution. This species holds a prominent position in biomedical research as one of the most extensively utilized experimental animals, and its whole-genome draft sequence was determined immediately after that of humans (Mouse Genome Sequencing Consortium 2002). The origin of *M. musculus* can be traced to South Asia, although its precise original location remains unclear (Boursot et al. 1993; Din et al. 1996). Earlier taxonomic studies based on morphology have proposed numerous subspecies (Fortuyn 1931; Marshall 1977; Schwarz and Schwarz 1943); however, currently, three principal subspecies, *M. m. castaneus*, *M. m. domesticus*, and *M. m. musculus*, have gained wide acknowledgment. The subspecies *castaneus* predominantly occupies South and Southeast Asia, showing remarkable genetic diversity relative to that of other subspecies. In contrast, the subspecies *domesticus* spans extensive areas in Western Europe, the Middle East, Africa, North and South America, Australia, and various remote islands. Meanwhile, the subspecies *musculus* can be found across northern Eurasia, including Eestern Europe, Central Asia, and Northeast Asia. Beyond these main subspecies, *M. m. bactrianus*, *M. m. gentilulus*, and *M. m. molossinus* are also well recognized, residing around the Iranian region, Yemen/Madagascar, and Japan, respectively. Nevertheless, genetic studies have indicated that the genetic characteristics of these subspecies are largely assigned to either a single feature or a combination of genetic features from the major subspecies (Fujiwara et al. 2022a, b; Lawal et al. 2022; Yonekawa et al. 1988). Figure 2.2 shows the results of the Principal Components Analysis (PCA) of house mouse samples worldwide analyzed by Fujiwara et al. (2022b), including the three major genetic components of wild house mice. However, these PCA results were highly dependent on the sample size; as such, even minor samples with highly differentiated genomes may not have been represented in the PCA plot. For example, a study by Fujiwara et al. (2022a), described in a later subsection, showed that Nepalese samples from Katmandu were somewhat differentiated from the other three subspecies.

Owing to their vital role in biomedical research, a multitude of extensive whole-genome sequencing projects have been conducted, and are still ongoing, for numerous inbred laboratory strains (Doran et al. 2016; Keane et al. 2011; Lilue et al. 2018; Takada et al. 2022). Genome-wide investigations of classical inbred strains have revealed that, although they predominantly consist of the genome of the subspecies *domesticus*, segments of their genetic material stem from the Japanese subspecies *molossinus* (Frazer et al. 2007; Takada et al. 2013; Yang et al. 2011). This pattern suggests that these laboratory strains originated from European fancy mice associated with *M. m. domesticus*, but also incorporated a minor proportion (less than 10%) of their genome from a Japanese fancy mouse strain imported from Japan to Europe as pets, which was subsequently examined in genome sequencing studies

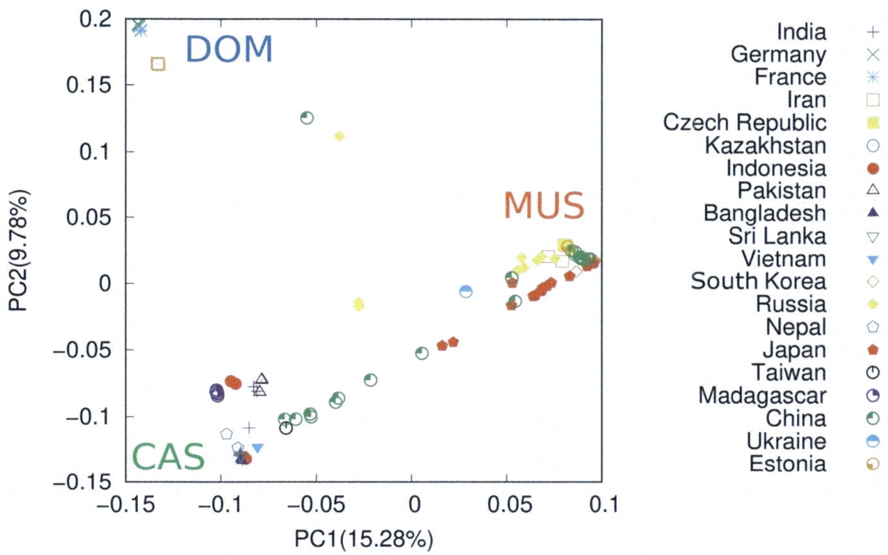

Fig. 2.2 PCA plot (*x*-axis: PC1; *y*-axis: PC2) depicting the global genetic diversity of *Mus musculus*. Each vertex of the triangle represents the genetic component of major three subspecies. (Adopted from Fujiwara et al. (2022b))

(Takada et al. 2013). In contrast, relatively few whole-genome sequencing studies have been conducted on wild house mice (Fujiwara et al. 2022a, b, 2024; Halligan et al. 2010; Harr et al. 2016; Lawal et al. 2022; Phifer-Rixey et al. 2018). These studies have repeatedly shown that the genetic diversity of classical inbred strains reflects only a minor portion of that in wild house mice (Salcedo et al. 2007).

This section aims to consolidate our understanding of the migration history of *M. musculus*, with a particular emphasis on East Asia, by drawing from genetic research. We began by exploring patterns in mitochondrial phylogenetic trees and then delved into the results of previous nuclear genome analyses. Collectively, these findings provide genetic evidence for the historical interactions between humans and *M. musculus*.

Phylogenetic Patterns of the House Mouse Mitochondrial Genome

The subspecies *musculus* originated in Northern Iran and was dispersed by prehistoric humans via western and eastern routes towards eastern Europe and East Asia, respectively. Analysis of mitochondrial genomes has suggested two eastern routes, one leading to Siberia and the Russian Far East, including Sakhalin, and the other leading to western and northeastern China (Suzuki et al. 2013). The home range of the subspecies *castaneus* spans the Middle East to northern India and Nepal. In prehistoric times, this species ranged from the eastern end near northeast India to southern China, and to both the continental and insular regions of Southeast Asia.

The availability of genome sequence data for more than 100 wild-derived house mice throughout Asia (Fujiwara et al. 2022a) has enabled phylogenetic analysis using mitochondrial genome sequences, facilitating the assessment of historical events and the reconstruction of prehistory with higher temporal and spatial resolution (Li et al. 2021). This study suggested that millet agriculture significantly diversified the subspecies *musculus* in northeastern China about 8000 years ago, and about 5000–6000 years ago, the subspecies *musculus* spread with millet farmers to the Korean Peninsula (Fig. 2.3).

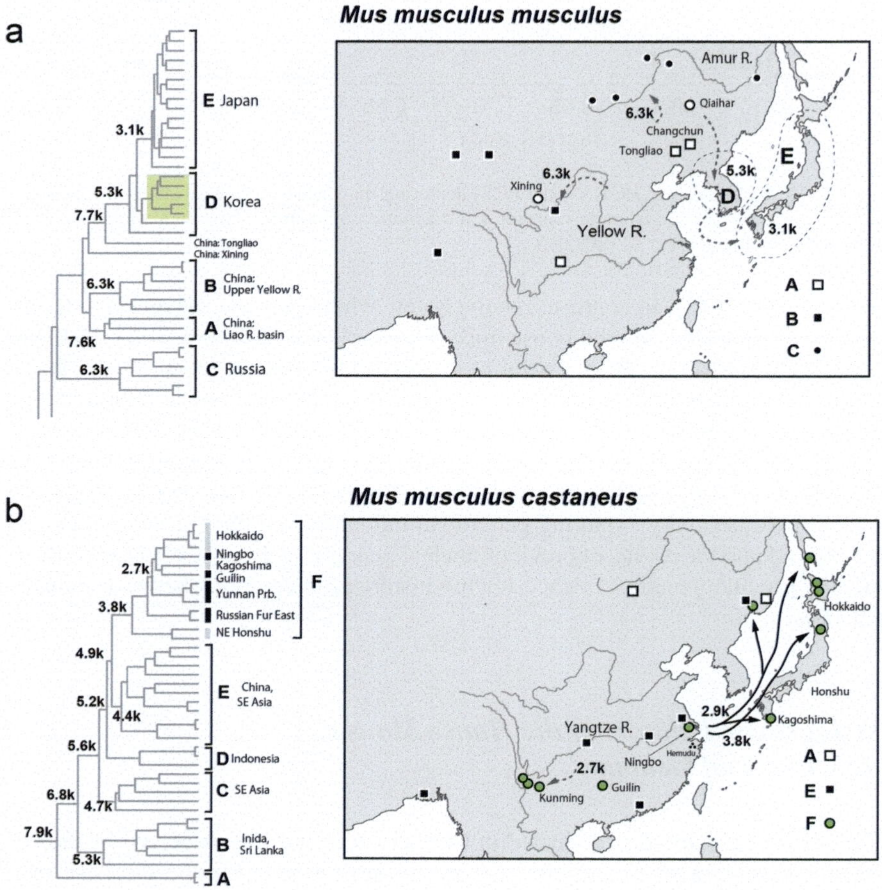

Fig. 2.3 Mitochondrial phylogenetic analysis of two subspecies groups, *Mus musculus musculus* (**a**) and *M. m. castaneus* (**b**), based on mitochondrial genomes. The phylogenetic trees and maps of the major haplotype groups, along with the numerical numbers on the trees and maps, represent estimated divergence and dispersal ages (×1000 years)

In contrast, the subspecies *castaneus* arrived in China approximately 8000 years ago, although the details of their arrival are unknown. However, approximately 4000 years ago, the diversification event of subspecies *castaneus* occurred in southern China, reaching the Japanese Archipelago approximately 3000 years ago (Fig. 2.3). The subspecies *castaneus* lineage extends to northeastern Japan, Kyushu, Sakhalin, and Primorye, Russia, suggesting transport via ships to these regions.

Genomic Patterns of Differentiation in House Mice

Fujiwara et al. (2022a) performed whole-genome sequencing analysis of nearly 100 house mouse genomes collected from a wide area of East Eurasia (Fujiwara et al. 2022a), which provided important information on the past demographics of mouse populations. Three subspecies showed distinct population demography in the last 200,000 years, assuming a mutation rate of 5.7×10^{-9} per generation (Milholland et al. 2017) and 1 year generation time. The subspecies *domesticus* and *musculus* showed strong population bottlenecks approximately 10,000 years ago, which was probably attributable to the founder effect associated with the spread of agriculture. Meanwhile, the bottleneck of the subspecies *castaneus* appeared earlier, and the reduction in population size was much milder than that of the other subspecies, probably because the extant distribution of the subspecies *castaneus* overlaps with its native range.

The analysis of nuclear genome sequences has been largely concordant with that of mitochondrial genome sequences; however, unlike mitochondrial genomes, nuclear genomes recombine with each other; thus, genomic introgression due to admixtures between subspecies is observable. The genetic components of East Asian house mouse genomes include the subspecies *castaneus* and *musculus*, and the differences among samples are mostly explained by different levels of admixture between the two subspecies.

Unlike the pattern of differentiation at the mitochondrial locus in the Japanese Archipelago, the majority (80% on average) of nuclear genomes are derived from the subspecies *musculus*, and the genetic component of the subspecies *musculus* is more pronounced on the Pacific Ocean side than on the Japanese Ocean side (Fujiwara et al. 2024). In the continental area, the genetic division of the subspecies *castaneus* and *musculus* occurred around the Yangtze River or Qinling Mountains. The genomes of southern Chinese populations include components of the subspecies *castaneus,* but have experienced a small amount of introgression from the subspecies *musculus*. Most samples from northern China, Korea, and Russia had strong genetic features shared with the subspecies *musculus*, except for some Russian samples around the Primorye region that contained the genetic component of the subspecies *domesticus*, which could be due to recent introgression through ports. Interestingly, among all samples they sequenced, Korean samples and Kazakhstani samples showed the lowest levels of introgression from the subspecies *castaneus*.

In addition to the inference of demographic history based on presumably neutral genetic markers, the pattern of differentiation of genes with phenotypic effects in East Asia has also been studied. Notably, the fur color-related genes *Asip* (Takeishi et al. 2021) and *Mc1r* (Zakoh et al. 2023) have been studied, revealing the associated haplotypes (~200 kb). These findings suggest that prehistoric migration plays a role in shaping the fur color variation exhibited by Japanese house mice. Specifically, the subspecies *musculus* from Europe and Russia displayed black ventral hair, whereas their counterparts from China, the Korean Peninsula, and most of Japan exhibited white ventral hair. The subspecies *castaneus* displays high dorsoventral lightness in all major regions of its homeland, especially in the Middle East. However, in the regions where they spread with humans during prehistoric times, they exhibit significantly darker dorsoventral and ventral characteristics. In addition to court color genes, Fujiwara et al. (2024) examined genomic loci under natural selection, during the admixture of the subspecies *musculus* and *castaneus* in the Japanese Archipelago. The analysis showed that although the genomic background of Japanese house mice was mostly derived from the subspecies *musculus*, some genes, particularly immune-related genes and vomeronasal receptor genes, exhibited *castaneus*-derived ancestry. These finding highlighted the importance of natural selection during the formation of hybrid genomes such as those of the Japanese house mice.

Migration Route of M. musculus *to the Japanese Archipelago*

Both mitochondrial and nuclear data clearly showed the dual structure of Japanese house mice, which are traditionally recognized as the subspecies *molossinus*. Mitochondrial phylogeny strongly implies that the subspecies *musculus* entered East Asia via the northern Himalayan route, paralleling Paleolithic human movements from central/western Eurasia to East Asia (Osada and Kawai 2021).

Archaeological findings highlight three primary Neolithic cultural hubs in East Asia: the Yellow River Basin and the West Liao River Basin, centered on foxtail and broomcorn millet farming, and the Yangtze River Basin, which is known for its rice cultivation. Recent sequencing of ancient human genomes has uncovered stronger genetic differentiation between the inhabitants of these regions than that observed in modern Chinese populations. The present distribution of subspecies *castaneus* and subspecies *musculus* mirrors the genetic patterns of these ancient humans, indicating a link to the agricultural practices of the time.

The current distribution of subspecies *musculus* in China corresponds to regions of dryland crop farming, such as foxtail and broomcorn millets, in the Early to Middle Neolithic periods. Therefore, it is reasonable to assume that millet farming contributed to the expansion of subspecies *musculus* in East Asia. In addition, recent archaeological and genetic studies have highlighted the important contribution of the West Liao River Basin to human colonization around northeastern China, the Korean Peninsula, and the Japanese Archipelago (Robbeets et al. 2021). The West

Liao River Basin is located at the base of the Korean Peninsula. Presently, this region is characterized by dry land, but it was conducive to millet farming during the Neolithic period (Li et al. 2020; Shelach-Lavi 2015). A recent study of seed impressions in pottery showed that early Yayoi migrants from the Korean Peninsula to the Japanese Archipelago in the Final Jomon and Initial Yayoi periods used both rice and millet, suggesting that the subspecies *musculus* may have arrived at the Japanese Archipelago through millet farming rather than rice farming.

The exact migration route of the subspecies *castaneus* from northern India to southern China remains ambiguous, though it has been linked to rice cultivation practices. Whether japonica rice (*Oryza sativa japonica*) and indica rice (*O. s. indica*) have been domesticated independently remains debated. Nevertheless, significant post-domestication gene flow between *japonica* and *indica* rice suggests active human interactions between these regions during the Early Neolithic period (Choi et al. 2017; Huang et al. 2012).

The arrival timeline of the subspecies *castaneus* in the Japanese Archipelago is also unclear. Analysis of mitochondrial phylogeny and whole-genome differentiation indicated that the subspecies *castaneus* did not accompany Yayoi migrants to Japan. Moreover, whether the subspecies *castaneus* entered the Japanese archipelago during the Jomon period is also not known; however, current genetic data show little association between the modern Japanese and Southern Chinese populations (Osada and Kawai 2021). Therefore, the subspecies *castaneus* may have reached the Japanese Archipelago through maritime trade rahter than associating with human migration. The precise timing of this migration is a subject for future research.

2.3 Rat

Overview

The genus *Rattus* encompasses up to 64 species (Thomson et al. 2018), yet the intricacies of their evolutionary relationships remain unclear. *Rattus* species can be predominantly found in Asian countries; however, species such as black and brown rats have expanded globally and live commensally with humans. Molecular phylogenetic studies using mitochondrial genomes estimated that the ancestors of these two groups diverged approximately 3 Mya (Robins et al. 2008).

The taxonomic classification of black rats has been a subject of debate, leading to their common reference as the *Rattus rattus* complex (RrC). In RrC, *R. rattus* and *R. tanezumi* are the two major groups. The terms black rat, house rat, and roof rat are generally used to refer to RrC as a whole, whereas Asian house rats or Tanezumi rats (Tanezumi, in Japanese, refers to paddy field rodents) specifically refer to the *R. tanezumi* group. *R. rattus* and *R. tanezumi* had distinct chromosome counts, with $2n = 38$ and 42 chromosomes, respectively. Despite these differences, which led some researchers to classify them as chromosomal species, their taxonomic relationships remain complex and unresolved (Aplin et al. 2003; Yosida et al. 1985). Notably, these groups have been observed to hybridize in areas where their habitats intersect, regardless of their chromosomal disparities (Conroy et al. 2013; Lack et al. 2012; Yosida et al. 1971). Similarly, the brown rat, also known as the sewer rat or Norway rat, has taxonomic ambiguities with its sister species, *Rattus nitidus* (Himalayan rat). A recent genome-wide sequencing study identified evidence of an admixture between brown and Himalayan rats in both mitochondrial and nuclear genomes (Teng et al. 2017). We present the phylogenetic tree of whole mitochondrial genomes encompassing brown and black rats reconstructed using unpublished and public datasets in Fig. 2.4. A detailed description of the tree is provided in the following section.

The spread of black rats to Europe was earlier than that of brown rats: around 100 BCE–200 CE and the 1700s for brown and black rats, respectively, although direct evidence is scarce (Hulme-Beaman et al. 2021); however, compared to that of house mice, how these *Rattus* species spread worldwide as commensal animals is largely unknown, and different studies have resulted in contradictory conclusions. Here, we summarize our current knowledge of genetics-based studies to clarify how black and brown rats have become two of the most common commensal animals in humans.

Rattus rattus *complex (RrC, black rats)*

A previous study conducted on the mtDNA analysis of RrC revealed distinct differences (Chinen et al. 2005). Analysis of many mtDNA lineages has shown that rat species distributed in Southeast Asia are closely related (Aplin et al. 2011; Conroy et al. 2013; Pagès et al. 2010; Robins et al. 2007; Tollenaere et al. 2010), and there are currently six closely related rat mtDNA lineages, consisting of RrC Lineages I to VI (Aplin et al. 2011). Samples of RrC Lineages I, II, IV, and VI are shown in Fig. 2.4. The three of six lineages exhibited a restricted geographic distribution. Specifically, RrC Lineage III was localized in Nepal and Pakistan, Lineage V in Vietnam and Thailand, and Lineage VI in Sumatra and the Malay Peninsula. In contrast, the remaining three showed rather wide ranges: RrC Lineage I was found ubiquitously worldwide, whereas Lineage VI extended from Myanmar to mainland Southeast Asia and eastern Asia, including Japan. Lineage IV is found primarily in the Malay Peninsula and the island regions of Southeast Asia, including Taiwan, the Philippines, and Indonesia.

Considering the results of mtDNA phylogenetic analyses (e.g., Kai et al. 2024), the ancestral lineage of RrC in Southeast Asia began to diverge approximately one million years ago, giving rise to RrC Lineages II, IV, and VI. Subsequently, the RrC Lineage II expanded into Nepal approximately 800,000 years ago (RrC Lineage III) and into the Indian subcontinent approximately 600,000 years ago (RrC Lineage I),

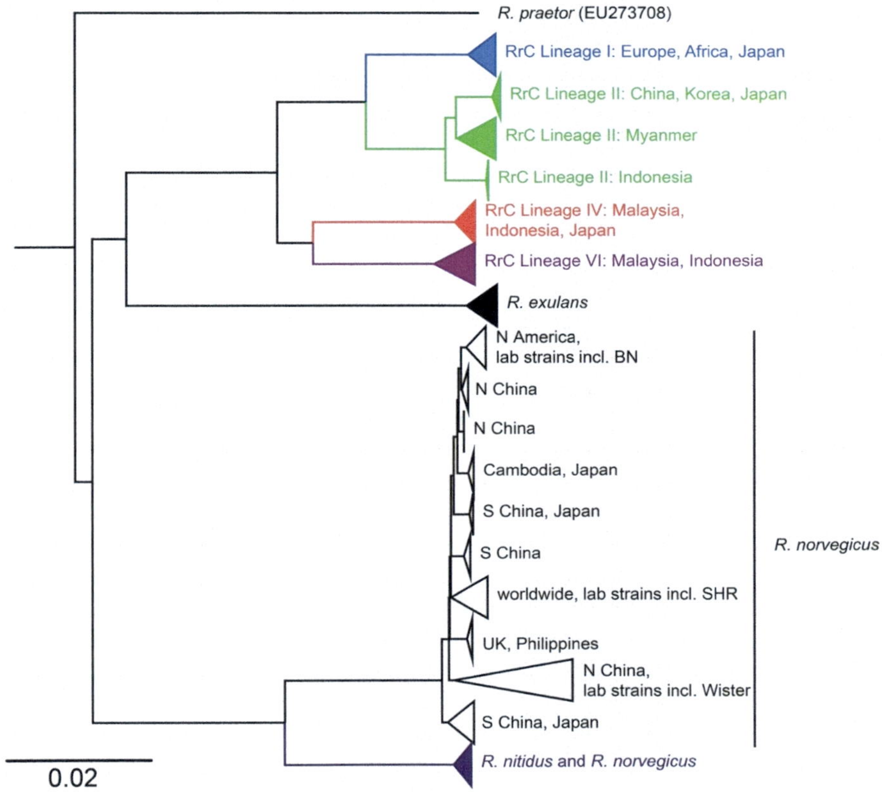

Fig. 2.4 The phylogenetic tree of whole mitochondrial genomes for various *Rattus* species, reconstructed using both unpublished and publicly available data. Mitochondrial genomes were assembled with GetOrganelle (Jin et al. 2020), and the tree was constructed using the Neighbor-joining method (Saitou and Nei 1987). A sample of *Mus musculus* was used as the outgroup, which is not depicted

giving rise to six mtDNA lineages. Subsequently, each mtDNA lineage underwent further subdivision, and the RrC Lineage II, for example, also appeared to have spread to Taiwan, possibly related to Pleistocene climate changes that occurred 250,000 years ago.

Three major RrC Lineages, I, II, and IV, evolved during the Holocene, and each appeared to have expanded into new regions with human agricultural activities. However, in prehistoric times, colonized areas may have been limited to those with subtropical conditions. The RrC Lineage II probably interacted with humans from southern China. The descendant lineages spread to the continental parts of Southeast Asia, including Vietnam and Thailand, and were further introduced to the Ryukyu Islands 1000–2000 years ago (Kai et al. 2024). In contrast, RrC Lineage IV underwent population expansion in the Malay Peninsula about 200,000 years ago, after which their range expanded to the islands in Southeast Asia, including the Philippines and Indonesia (Kai et al. 2024).

In the Japanese Archipelago, the introduction of black rats to Honshu and Hokkaido is thought to have been more recent than that in the Ryukyu Islands. Warm environments are important for the establishment of uninhabited subtropical islands, whereas anthropogenic warming is important for colder regions. In the subtropical zone, the introduction of black rats is easy; thus, in the Japanese Archipelago, RrC Lineage IV is distributed in the Miyako Islands, and its introduction is considered to be a recent event.

Genetic analysis of black rats based on nuclear genome sequences is still in progress, and results are scarce. In particular, the genetic differentiation of individuals with different mitochondrial lineages has only been partially analyzed using ddRAD-seq data (Thomson et al. 2022). The results showed genetic similarities between RrC Lineages II and IV in the nuclear genome, suggesting that mitochondrial lineages do not fully represent the genetic characteristics of RrC populations. In addition, many samples showed incongruent genetic patterns between the nuclear and mitochondrial loci (Thomson et al. 2022), suggesting that the mitochondrial lineages of the samples may not be appropriate criteria for species/subspecies/population delimitation. Delimitation is particularly challenging because RrC shows extremely high morphological variation within populations, and the current classification of species mostly relies on genetic evidence such as chromosome numbers and mitochondrial haplotypes; such genetic analysis has not been thoroughly performed. Consequently, there is a misconception that RrC Lineage IV generally has a karyotype of $n = 40$. Further population genomics studies are required to define the pattern of genetic differentiation within RrC.

Although some inconsistencies exist between the mitochondrial and nuclear genome phylogenies, there is a consensus that black rat (RrC Lineage I) populations have expanded from the Indian regions to Europe. Whole genome sequencing studies have suggested that the two primary populations, identified as *R. rattus* and *R. tanezumi*, diverged approximately 120,000 years ago, each following a distinct population trajectory (Yu et al. 2022). The *R. rattus* population displayed evidence of a significant population bottleneck after divergence, unlike the *R. tanezumi* population, suggesting that *R. rattus*, which possesses a different chromosome number, was a derived population.

Black rats are characterized by melanistic fur color variation, primarily in RrC Lineage I (Fig. 2.1), though they are thought to have spread owing to anthropogenic dispersal, observed not only in Europe and America, but also in Oceania and Yaponesia. The responsible mutation has been identified as a single amino acid mutation in the *Mc1r* gene (Kambe et al. 2011). However, melanism can occur spontaneously in any region, as evidenced by observations from Okinawa Island. This mutation has been attributed to the loss of one cysteine residue in the *Asip* gene in RrC Lineage II rats (Sasamori et al. 2017).

Rattus norvegicus *(brown rats)*

Given their significance in biomedical research, many studies have assessed the genome-wide diversity of brown rats (Deinum et al. 2015; Jing et al. 2022; Puckett et al. 2016, 2018; Puckett and Munshi-South 2019; Teng et al. 2017; Zeng et al. 2018). These studies revealed that brown rats exhibit significantly less genetic diversity than do black rats, even among wild populations. Asian brown rat populations are genetically divided into two major groups: southeastern and northeastern Asian populations (Puckett and Munshi-South 2019; Zeng et al. 2018). However, the debate continues regarding which group is the ancestral population. Traditional belief suggests that the brown rat's native habitat is in northern or central Asia, including Mongolia, but a recent whole-genome sequencing study propose a Southeast Asian origin (Zeng et al. 2018). An alternative hypothesis suggests an unknown ancestral population in northern Asia, with the northern population diverging first, then the southern population. This model is supported by the presence of highly divergent populations in the Aleutian Islands and western North America (Puckett and Munshi-South 2019). Despite this controversy, most genetic evidence supports the idea that European and North American populations originated from the southern population rather than from the northern population. Since their arrival in Europe, various laboratory species of brown rats have been established in European and North American populations.

Whole Mitochondrial Genome Analysis of the Genus Rattus

In Fig. 2.4, we present a mitochondrial phylogenetic tree of black rats, brown rats, and other closely related rat species constructed using both publicly available data and new, unpublished datasets. This tree encompasses four *Rattus* complex/species: RrC, *Rattus exulans*, *R. norvegicus*, and *R. nitidus*. All groups except *R. nitidus* demonstrated monophyletic relationships within their respective categories.

Within RrC, our samples included RrC Lineages I, II, IV, and VI. The Japanese Archipelago samples fell into RrC Lineages I, II, and IV, of which Lineage II was predominant. RrC Lineage I samples were exclusive to Hokkaido and Ogasawara Islands, whereas RrC Lineage IV samples were exclusive to the Ryukyu Islands. The Japanese RrC Lineage II samples further bifurcated into mainland and Ryukyu clusters. The mainland cluster was closely related to a sample from Jeju Island, which was probably mislabeled as a brown rat (DDBJ/EMBL/GenBank accession number KF011916). This cluster is similar to that of another misidentified sample from China. These patterns suggest a relatively recent migration of RrC Lineage II black rats to Japan and a recent admixture on the Ryukyu Islands.

In contrast to black rats, the mitochondrial phylogeny of brown rats showed little geographic structure. Japanese brown rats appeared only sporadically in the phylogenetic tree. Their branches were notably longer than those of the other species, suggesting ancestral polymorphisms in native brown rat populations. Alternatively, they may have been introduced to the Japanese Archipelago multiple times, and either the source populations were extinct or were not included in our dataset. Further analysis of nuclear genomes could shed light on the timing and pathways of the brown rat introduction to the Japanese Archipelago. The markedly different mitochondrial phylogenetic patterns observed in Japanese black and brown rats suggest that the mechanisms and timelines of their introduction into the archipelago vary significantly.

2.4 Perspectives

Here we provide a concise overview of the history of three major commensal rodent groups, as revealed by genetic studies. Despite being well recognized and extensively studied in laboratory settings, the natural history and genetic background of these species remain surprisingly enigmatic. Issues such as species delimitation, particularly in RrC, require more in-depth genetic research for a clearer understanding.

Advancements in methods such as ancient DNA sequencing from faunal remains and analysis of museum specimens have significantly enriched our knowledge of how these rodents have adapted to human-influenced environments (Yu et al. 2022). Continued sequencing of wild-derived samples using increasingly affordable sequencing technologies is essential to fully unravel their historical narratives.

Whole-genome sequencing is key to understanding the links between phenotypes and genotypes and how genetic changes have driven species' adaptations to their environment. In particular, house mice and brown rats are valuable for studying how natural variation influences adaptive phenotypes in controlled experiments. Ongoing large-scale studies on the phenotypic variations in these species are contributing to a growing database, enabling researchers to correlate specific genes with phenotypes. Identifying mutations that have directly influenced adaptive phenotypic changes during commensalism to humans, especially over the last 10,000 years, will enhance our understanding of the co-evolutionary dynamics between humans and these commensal animals.

Acknowledgements We express our gratitude to Takanori Mogi for the assistance in assembling the complete mitochondrial DNA sequences. We thank the members of the Yaponesian Genome Project. This work was supported by MEXT KAKENHI (grants 18H05511 and 23H04846 to Naoki Osada and grant 18H05508 to Hitoshi Suzuki).

Competing Interests The authors have no relevant financial or non-financial interests to disclose.

References

Aplin K, Brown P, Jacob J, Krebs C, Singleton G (2003) Field methods for rodent studies in Asia and the Indo-Pacific. ACIAR Monograph No. 100, p 223

Aplin KP, Suzuki H, Chinen AA, Chesser RT, ten Have J, Donnellan SC, Austin J, Frost A, Gonzalez JP, Herbreteau V, Catzeflis F, Soubrier J, Fang YP, Robins J, Matisoo-Smith E, Bastos ADS, Maryanto I, Sinaga MH, Denys C, Cooper A (2011) Multiple geographic origins of commensalism and complex dispersal history of black rats. PLoS One 6(11):e26357. https://doi.org/10.1371/journal.pone.0026357

Bae CJ, Douka K, Petraglia MD (2017) On the origin of modern humans: Asian perspectives. Science 358(6368):eaai9067. https://doi.org/10.1126/science.aai9067

Boursot P, Auffray JC, Britton-Davidian J, Bonhomme F (1993) The evolution of house mice. Annu Rev Ecol Syst 24(1):119–152. https://doi.org/10.1146/annurev.es.24.110193.001003

Chinen AA, Suzuki H, Aplin KP, Tsuchiya K, Suzuki S (2005) Preliminary genetic characterization of two lineages of black rats (*Rattus rattus* sensu lato) in Japan, with evidence for introgression at several localities. Genes Genet Syst 80(5):367–375. https://doi.org/10.1266/ggs.80.367

Choi JY, Platts AE, Fuller DQ, Hsing YI, Wing RA, Purugganan MD (2017) The rice paradox: multiple origins but single domestication in Asian rice. Mol Biol Evol 34(4):969–979. https://doi.org/10.1093/molbev/msx049

Conroy CJ, Rowe KC, KMC R, Kamath PL, Aplin KP, Hui L, James DK, Moritz C, Patton JL (2013) Cryptic genetic diversity in *Rattus* of the San Francisco Bay region, California. Biol Invasions 15(4):741–758. https://doi.org/10.1007/s10530-012-0323-9

de Jong TV, Chen H, Brashear WA, Kochan KJ, Hillhouse AE, Zhu Y, Dhande IS, Hudson EA, Sumlut MH, Smith ML, Kalbfleisch TS, Doris PA (2022) mRatBN7.2: familiar and unfamiliar features of a new rat genome reference assembly. Physiol Genomics 54(7):251–260. https://doi.org/10.1152/physiolgenomics.00017.2022

Deinum EE, Halligan DL, Ness RW, Zhang YH, Cong L, Zhang JX, Keightley PD (2015) Recent evolution in *Rattus norvegicus* is shaped by declining effective population size. Mol Biol Evol 32(10):2547–2558. https://doi.org/10.1093/molbev/msv126

Din W, Anand R, Boursot P, Darviche D, Dod B, Jouvin-Marche E, Orth A, Talwar GP, Cazenave PA, Bonhomme F (1996) Origin and radiation of the house mouse: clues from nuclear genes. J Evol Biol 9(5):519–539. https://doi.org/10.1046/j.1420-9101.1996.9050519.x

Doran AG, Wong K, Flint J, Adams DJ, Hunter KW, Keane TM (2016) Deep genome sequencing and variation analysis of 13 inbred mouse strains defines candidate phenotypic alleles, private variation and homozygous truncating mutations. Genome Biol 17(1):167. https://doi.org/10.1186/s13059-016-1024-y

Douglass K, Gaffney D, Feo TJ, Bulathsinhala P, Mack AL, Spitzer M, Summerhayes GR (2021) Late Pleistocene/Early Holocene sites in the montane forests of New Guinea yield early record of cassowary hunting and egg harvesting. Proc Natl Acad Sci U S A 118(40):e2100117118. https://doi.org/10.1073/pnas.2100117118

Fortuyn ABD (1931) *Mus musculus* and *Mus wangeri wagneri* compared. I. the number of tail-rings. Genetics 16(2):160–167

Frazer KA, Eskin E, Kang HM, Bogue MA, Hinds DA, Beilharz EJ, Gupta RV, Montgomery J, Morenzoni MM, Nilsen GB, Pethiyagoda CL, Stuve LL, Johnson FM, Daly MJ, Wade CM, Cox DR (2007) A sequence-based variation map of 8.27 million SNPs in inbred mouse strains. Nature 448(7157):1050–1053. https://doi.org/10.1038/nature06067

Fujiwara K, Kawai Y, Takada T, Shiroishi T, Saitou N, Suzuki H, Osada N (2022a) Insights into *Mus musculus* population structure across Eurasia revealed by whole-genome analysis. Genome Biol Evol 14(5):evac068. https://doi.org/10.1093/gbe/evac068

Fujiwara K, Ranorosoa MC, Ohdachi SD, Arai S, Sakuma Y, Suzuki H, Osada N (2022b) Whole-genome sequencing analysis of wild house mice (*Mus musculus*) captured in Madagascar. Genes Genet Syst 97(4):193–207. https://doi.org/10.1266/ggs.22-00090

Fujiwara K, Kubo S, Endo T, Takada T, Shiroishi T, Suzuki H, Osada N (2024) Inference of selective forces on house mouse genomes during secondary contact in East Asia. Genome Res 34(3):366–375. https://doi.org/10.1101/gr.278828.123

Halligan DL, Oliver F, Eyre-Walker A, Harr B, Keightley PD (2010) Evidence for pervasive adaptive protein evolution in wild mice. PLoS Genet 6(1):e1000825. https://doi.org/10.1371/journal.pgen.1000825

Harr B, Karakoc E, Neme R, Teschke M, Pfeifle C, Pezer Ž, Babiker H, Linnenbrink M, Montero I, Scavetta R, Abai MR, Molins MP, Schlegel M, Ulrich RG, Altmüller J, Franitza M, Büntge A, Künzel S, Tautz D (2016) Genomic resources for wild populations of the house mouse, *Mus musculus* and its close relative *Mus spretus*. Sci Data 3:160075. https://doi.org/10.1038/sdata.2016.75

Huang X, Kurata N, Wei X, Wang ZX, Wang A, Zhao Q, Zhao Y, Liu K, Lu H, Li W, Guo Y, Lu Y, Zhou C, Fan D, Weng Q, Zhu C, Huang T, Zhang L, Wang Y, Han B (2012) A map of rice genome variation reveals the origin of cultivated rice. Nature 490(7421):497–501. https://doi.org/10.1038/nature11532

Hulme-Beaman A, Dobney K, Cucchi T, Searle JB (2016) An ecological and evolutionary framework for commensalism in anthropogenic environments. Trends Ecol Evol 31(8):633–645. https://doi.org/10.1016/j.tree.2016.05.001

Hulme-Beaman A, Orton D, Cucchi T (2021) The origins of the domesticate brown rat (*Rattus norvegicus*) and its pathways to domestication. Anim Front 11(3):78–86. https://doi.org/10.1093/af/vfab020

Jin JJ, Yu WB, Yang JB, Song Y, dePamphilis CW, Yi TS, Li DZ (2020) GetOrganelle: a fast and versatile toolkit for accurate de novo assembly of organelle genomes. Genome Biol 21(1):241. https://doi.org/10.1186/s13059-020-02154-5

Jing M, Chen Y, Yao K, Wang Y, Huang L (2022) Comparative phylogeography of two commensal rat species (*Rattus tanezumi* and *Rattus norvegicus*) in China: insights from mitochondrial DNA, microsatellite, and 2b-RAD data. Ecol Evol 12(10):e9409. https://doi.org/10.1002/ece3.9409

Jones EP, Eager HM, Gabriel SI, Jóhannesdóttir F, Searle JB (2013) Genetic tracking of mice and other bioproxies to infer human history. Trends Genet 29(5):298–308. https://doi.org/10.1016/j.tig.2012.11.011

Kai H, Takda N, Thomson V, Suzuki H (2024) Region-specific genetic diversity of black rats (*Rattus rattus* complex) in Southeast and East Asia shaped by rapid population expansion events. Zoolog Sci 41(3):290–301

Kambe Y, Tanikawa T, Matsumoto Y, Tomozawa M, Aplin KP, Suzuki H (2011) Origin of agouti-melanistic polymorphism in wild black rats (*Rattus rattus*) inferred from *Mc1r* gene sequences. Zoolog Sci 28(8):560–567., 568. https://doi.org/10.2108/zsj.28.560

Kawamura Y (1988) Quaternary rodent faunas in the Japanese Islands (Part 1). Mem Fac Sci Kyoto Univ Ser Geol Mineral 53(1–2):331–348

Kawamura Y (1989) Quaternary rodent faunas in the Japanese Islands (Part 2). Mem Fac Sci Kyoto Univ Ser Geol Mineral 54(1–2):1–235

Keane TM, Goodstadt L, Danecek P, White MA, Wong K, Yalcin B, Heger A, Agam A, Slater G, Goodson M, Furlotte NA, Eskin E, Nellåker C, Whitley H, Cleak J, Janowitz D, Hernandez-Pliego P, Edwards A, Belgard TG, Adams DJ (2011) Mouse genomic variation and its effect on phenotypes and gene regulation. Nature 477(7364):289–294. https://doi.org/10.1038/nature10413

Kumar S, Stecher G, Suleski M, Hedges SB (2017) TimeTree: a resource for timelines, timetrees, and divergence times. Mol Biol Evol 34(7):1812–1819. https://doi.org/10.1093/molbev/msx116

Kuramoto T (2011) *Yoso-Tama-No-Kakehash*; The first Japanese guidebook on raising rats. Exp Anim 60(1):1–6. https://doi.org/10.1538/expanim.60.1

Lack JB, Greene DU, Conroy CJ, Hamilton MJ, Braun JK, Mares MA, Van Den Bussche RA (2012) Invasion facilitates hybridization with introgression in the *Rattus rattus* species complex. Mol Ecol 21(14):3545–3561. https://doi.org/10.1111/j.1365-294X.2012.05620.x

Lawal RA, Mathis VL, Barter ME, Charette JR, Garretson A, Dumont BL (2022) Taxonomic assessment of two wild house mouse subspecies using whole-genome sequencing. Sci Rep 12(1):20866. https://doi.org/10.1038/s41598-022-25420-x

Li T, Ning C, Zhushchikhovskaya IS, Hudson MJ, Robbeets M (2020) Millet agriculture dispersed from Northeast China to the Russian Far East: integrating archaeology, genetics, and linguistics. Archaeol Res Asia 22:100177. https://doi.org/10.1016/j.ara.2020.100177

Li Y, Fujiwara K, Osada N, Kawai Y, Takada T, Kryukov AP, Abe K, Yonekawa H, Shiroishi T, Moriwaki K, Saitou N, Suzuki H (2021) House mouse *Mus musculus* dispersal in East Eurasia inferred from 98 newly determined complete mitochondrial genome sequences. Heredity 126(1):132–147. https://doi.org/10.1038/s41437-020-00364-y

Lilue J, Doran AG, Fiddes IT, Abrudan M, Armstrong J, Bennett R, Chow W, Collins J, Collins S, Czechanski A, Danecek P, Diekhans M, Dolle DD, Dunn M, Durbin R, Earl D, Ferguson-Smith A, Flicek P, Flint J, Keane TM (2018) Sixteen diverse laboratory mouse reference genomes define strain-specific haplotypes and novel functional loci. Nat Genet 50(11):1574–1583. https://doi.org/10.1038/s41588-018-0223-8

Marshall JT Jr (1977) A synopsis of Asian species of *Mus* (Rodentia, Muridae). Bull Am Mus Nat Hist 158:173–220. https://cir.nii.ac.jp/crid/1573950400012850304

Milholland B, Dong X, Zhang L, Hao X, Suh Y, Vijg J (2017) Differences between germline and somatic mutation rates in humans and mice. Nat Commun 8(1):15183. https://doi.org/10.1038/ncomms15183

Mouse Genome Sequencing Consortuim (2002) Initial sequencing and comparative analysis of the mouse genome. Nature 420(6915):520–562. https://doi.org/10.1038/nature01262

Osada N, Kawai Y (2021) Exploring models of human migration to the Japanese archipelago using genome-wide genetic data. Anthropol Sci 129(1):45–58. https://doi.org/10.1537/ase.201215

Pagès M, Chaval Y, Herbreteau V, Waengsothorn S, Cosson JF, Hugot JP, Morand S, Michaux J (2010) Revisiting the taxonomy of the Rattini tribe: a phylogeny-based delimitation of species boundaries. BMC Evol Biol 10(1):184. https://doi.org/10.1186/1471-2148-10-184

Phifer-Rixey M, Bi K, Ferris KG, Sheehan MJ, Lin D, Mack KL, Keeble SM, Suzuki TA, Good JM, Nachman MW (2018) The genomic basis of environmental adaptation in house mice. PLoS Genet 14(9):e1007672. https://doi.org/10.1371/journal.pgen.1007672

Puckett EE, Munshi-South J (2019) Brown rat demography reveals pre-commensal structure in eastern Asia before expansion into Southeast Asia. Genome Res 29(5):762–770. https://doi.org/10.1101/gr.235754.118

Puckett EE, Park J, Combs M, Blum MJ, Bryant JE, Caccone A, Costa F, Deinum EE, Esther A, Himsworth CG, Keightley PD, Ko A, Lundkvist Å, McElhinney LM, Morand S, Robins J, Russell J, Strand TM, Suarez O, Munshi-South J (2016) Global population divergence and admixture of the brown rat (*Rattus norvegicus*). Proc R Soc B Biol Sci 283(1841):20161762. https://doi.org/10.1098/rspb.2016.1762

Puckett EE, Micci-Smith O, Munshi-South J (2018) Genomic analyses identify multiple Asian origins and deeply diverged mitochondrial clades in inbred brown rats (*Rattus norvegicus*). Evol Appl 11(5):718–726. https://doi.org/10.1111/eva.12572

Rane R, Pichlmueler F, Lee R, Edwards O, Gemmell N, Tompkins D (2020) *Rattus rattus* chromosomal assembly (New Zealand strain). v2. CSIRO. Data Collection. https://doi.org/10.25919/5e4f60ea8b64e

Robbeets M, Bouckaert R, Conte M, Savelyev A, Li T, An DI, Shinoda K, Cui Y, Kawashima T, Kim G, Uchiyama J, Dolińska J, Oskolskaya S, Yamano KY, Seguchi N, Tomita H, Takamiya H, Kanzawa-Kiriyama H, Oota H, Ning C (2021) Triangulation supports agricultural spread of the Transeurasian languages. Nature 599(7886):616–621. https://doi.org/10.1038/s41586-021-04108-8

Robins JH, Hingston M, Matisoo-Smith E, Ross HA (2007) Identifying Rattus species using mitochondrial DNA. Mol Ecol Notes 7(5):717–729. https://doi.org/10.1111/j.1471-8286.2007.01752.x

Robins JH, McLenachan PA, Phillips MJ, Craig L, Ross HA, Matisoo-Smith E (2008) Dating of divergences within the *Rattus* genus phylogeny using whole mitochondrial genomes. Mol Phylogenet Evol 49(2):460–466. https://doi.org/10.1016/j.ympev.2008.08.001

Saitou N, Nei M (1987) The neighbor-joining method: a new method for reconstructing phylogenetic trees. Mol Biol Evol 4:406–425. http://mbe.oxfordjournals.org/cgi/reprint/4/4/406.pdf

Salcedo T, Geraldes A, Nachman MW (2007) Nucleotide variation in wild and inbred mice. Genetics 177(4):2277–2291. https://doi.org/10.1534/genetics.107.079988

Sasamori S, Wiewel AS, Thomson VA, Kobayashi M, Nakata K, Suzuki H (2017) Potential causative mutation for melanism in rats identified in the Agouti signaling protein Gene (*Asip*) of the *Rattus rattus* Species Complex on Okinawa Island, Japan. Zoolog Sci 34(6):513–522. https://doi.org/10.2108/zs170027

Schwarz E, Schwarz HK (1943) The wild and commensal stocks of the house mouse, *Mus musculus* Linnaeus. J Mammal 24(1):59–72. https://doi.org/10.2307/1374781

Serikawa T (2004) Colourful history of Japan's rat resources. Nature 429(6987):15–15. https://doi.org/10.1038/429015b

Shelach-Lavi G (2015) The archaeology of early China: from prehistory to the Han Dynasty. Cambridge University Press, Cambridge. https://doi.org/10.1017/CBO9781139022682

Suzuki H, Nunome M, Kinoshita G, Aplin KP, Vogel P, Kryukov AP, Jin ML, Han SH, Maryanto I, Tsuchiya K, Ikeda H, Shiroishi T, Yonekawa H, Moriwaki K (2013) Evolutionary and dispersal history of Eurasian house mice *Mus musculus* clarified by more extensive geographic sampling of mitochondrial DNA. Heredity 111(5):375–390. https://doi.org/10.1038/hdy.2013.60

Takada T, Ebata T, Noguchi H, Keane TM, Adams DJ, Narita T, Shin-I T, Fujisawa H, Toyoda A, Abe K, Obata Y, Sakaki Y, Moriwaki K, Fujiyama A, Kohara Y, Shiroishi T (2013) The ancestor of extant Japanese fancy mice contributed to the mosaic genomes of classical inbred strains. Genome Res 23(8):1329–1338. https://genome.cshlp.org/content/23/8/1329

Takada T, Fukuta K, Usuda D, Kushida T, Kondo S, Kawamoto S, Yoshiki A, Obata Y, Fujiyama A, Toyoda A, Noguchi H, Shiroishi T, Masuya H (2022) MoG+: a database of genomic variations across three mouse subspecies for biomedical research. Mamm Genome 33(1):31–43. https://doi.org/10.1007/s00335-021-09933-w

Takeishi T, Fujiwara K, Osada N, Mita A, Takada T, Shiroishi T, Suzuki H (2021) Phylogeographic study using nuclear genome sequences of *Asip* to infer the origins of ventral fur color variation in the house mouse *Mus musculus*. Genes Genet Syst 96(6):271–284. https://doi.org/10.1266/ggs.21-00075

Teng H, Zhang Y, Shi C, Mao F, Hou L, Guo H, Sun Z, Zhang J (2016) Whole-genome sequencing reveals genetic variation in the Asian house rat. G3 (Bethesda) 6(7):1969–1977. https://doi.org/10.1534/g3.116.029504

Teng H, Zhang Y, Shi C, Mao F, Cai W, Lu L, Zhao F, Sun Z, Zhang J (2017) Population genomics reveals speciation and introgression between brown Norway rats and their sibling species. Mol Biol Evol 34(9):2214–2228. https://doi.org/10.1093/molbev/msx157

Thomson V, Wiewel A, Chinen A, Maryanto I, Sinaga MH, How R, Aplin K, Suzuki H (2018) A perspective for resolving the systematics of *Rattus*, the vertebrates with the most influence on human welfare. Zootaxa 4459(3):431–452. https://doi.org/10.11646/zootaxa.4459.3.2

Thomson VA, Wiewel AS, Palmer R, Hamilton N, Algar D, Pink C, Mills H, Aplin KP, Clark G, Anderson A, Herrera MB, Myers S, Bertozzi T, Piper PJ, Suzuki H, Donnellan S (2022) Genetic insights into the introduction history of black rats into the Eastern Indian ocean. Front Ecol Evol 9:786510. https://www.frontiersin.org/articles/10.3389/fevo.2021.786510

Tollenaere C, Brouat C, Duplantier JM, Rahalison L, Rahelinirina S, Pascal M, Moné H, Mouahid G, Leirs H, Cosson J-F (2010) Phylogeography of the introduced species *Rattus rattus* in the western Indian Ocean, with special emphasis on the colonization history of Madagascar. J Biogeogr 37(3):398–410. https://doi.org/10.1111/j.1365-2699.2009.02228.x

Yang H, Wang JR, Didion JP, Buus RJ, Bell TA, Welsh CE, Bonhomme F, Yu AHT, Nachman MW, Pialek J, Tucker P, Boursot P, McMillan L, Churchill GA, de Villena FPM (2011) Subspecific origin and haplotype diversity in the laboratory mouse. Nat Genet 43(7):648–655. https://doi.org/10.1038/ng.847

Yonekawa H, Moriwaki K, Gotoh O, Miyashita N, Matsushima Y, Shi LM, Cho WS, Zhen XL, Tagashira Y (1988) Hybrid origin of Japanese mice "*Mus musculus molossinus*": evidence from restriction analysis of mitochondrial DNA. Mol Biol Evol 5(1):63–78. https://doi.org/10.1093/oxfordjournals.molbev.a040476

Yosida TH, Kato H, Tsuchiya K, Moriwaki K (1971) Karyotypes and serum transferrin patterns of hybrids between Asian and Oceanian black rats, *Rattus rattus*. Chromosoma 34(1):40–50. https://doi.org/10.1007/BF00285515

Yosida TH, Udagawa T, Ishibashi M, Moriwaki K, Yabe T, Hamada T (1985) Studies on the Karyotypes of the Black Rats distributed in the Pacific and South Pacific Islands, with special regard to the border line of the Asian and Oceanian type black rats on the Pacific ocean. Proc Jpn Acad Ser B 61(2):71–74. https://doi.org/10.2183/pjab.61.71

Yu H, Jamieson A, Hulme-Beaman A, Conroy CJ, Knight B, Speller C, Al-Jarah H, Eager H, Trinks A, Adikari G, Baron H, Böhlendorf-Arslan B, Bohingamuwa W, Crowther A, Cucchi T, Esser K, Fleisher J, Gidney L, Gladilina E, Orton D (2022) Palaeogenomic analysis of black rat (*Rattus rattus*) reveals multiple European introductions associated with human economic history. Nat Commun 13(1):2399. https://doi.org/10.1038/s41467-022-30009-z

Zakoh K, Fujiwara K, Takada T, Osada N, Suzuki H (2023) Genealogical characterization of regional populations and dorsal coat color variation in the house mouse *Mus musculus* from Asia based on haplotype structure analysis of a gene-rich region harboring *Mc1r*. Genes Genet Syst 98(2):73–87. https://doi.org/10.1266/ggs.22-00157

Zeng L, Ming C, Li Y, Su LY, Su YH, Otecko NO, Dalecky A, Donnellan S, Aplin K, Liu XH, Song Y, Zhang ZB, Esmailizadeh A, Sohrabi SS, Nanaei HA, Liu H, Wang MS, Ag Atteynine S, Rocamora G, Zhang YP (2018) Out of southern East Asia of the brown rat revealed by large-scale genome sequencing. Mol Biol Evol 35(1):149–158. https://doi.org/10.1093/molbev/msx276

Chapter 3
Phylogeography of the Brown Bear (*Ursus arctos*) and Bear-Human Relationships in the Far East, Including Hokkaido Island, Japan

Ryuichi Masuda

Abstract This chapter discusses the phylogeography of the brown bear and bear-human relationships in the Far East, including Japan, based on genetic data from our lab. The maternally inherited mitochondrial DNA (mtDNA) phylogeography of brown bears in the Northern Hemisphere has been studied for the past 30 years. Given the female philopatry of this species, geographic information on mtDNA distribution provides invaluable insights into the migration history of the bears across their wide range. The brown bear population on Hokkaido Island, northern Japan, comprises three distinct mtDNA lineages, which can be called a "triple mtDNA population structure", compared with those of the Eurasian and American continents. By contrast, an analysis of paternally inherited Y-chromosomal DNA from male bears showed no evidence of intra-population structure on Hokkaido, like that of the triple structure identified from mtDNA analyses. This difference might have arisen through male-mediated gene flow among local natal populations, combined with founder effects and subsequent genetic drift after the bears' immigration to Hokkaido from the continent. In addition, a whole-genomic analysis revealed that the Hokkaido bear population formed a different clade from those in Europe and North America, suggesting that it maintained some genetic diversity even after geographical isolation on the island. The genetic uniformity within the Hokkaido population could have resulted from male-mediated gene flow, and this is congruent with the Y-chromosomal DNA data. Genetic data on brown bears can sometimes be useful for studies of the bear-human relationships that have developed in indigenous cultures. For example, the *iomante* bear-sending ritual performed in the Ainu Culture on Hokkaido involves juvenile bears, and knowing its origin is very important for understanding cultural history and development in the context of archaeology and ethnology. Analyses of ancient mtDNA from bear remains excavated from the Okhotsk Cultural archaeological site on Rebun Island, northern

R. Masuda (✉)
Department of Biological Sciences, Faculty of Science, Hokkaido University, Sapporo, Japan
e-mail: masudary@sci.hokudai.ac.jp

N. Osada et al. (eds.), *Phylogeographic History of Plants and Animals Coexisting with Humans in Asia*, Evolutionary Studies,
https://doi.org/10.1007/978-981-97-6887-5_3

Hokkaido, clarified that the *iomante* ritual could have originated in the Okhotsk Culture, and that people in different ancient cultures had similarly imparted significance to and placed high value on juvenile brown bears, which functioned in intercultural communication and bonding.

Keywords Bear-sending ritual · Brown bear · Hokkaido · Northern Hemisphere · Phylogeography

Abbreviation

mtDNA mitochondrial DNA

3.1 Introduction

In this chapter, based on genetic data from our lab, I will discuss the phylogeography of the brown bear, which is widely distributed in the Northern Hemisphere, and bear-human relationships in the Far East, including Japan.

The topic of this section is the general biology of the brown bear (*Ursus arctos*). This species is the largest terrestrial carnivore (order Carnivora), and generally omnivorous in a variety of habitats. The brown bear has adapted to a wide range of environments, mainly forests in subarctic Eurasia and North America. In the Japanese archipelago, it presently occurs only on Hokkaido Island, the northernmost main island.

The wide distribution of the brown bear has led to geographical variation in morphology among local populations, including in such features as coat color and body size. Bears in colder parts of the range are generally larger than those of the same sex in warmer areas, a trend in body size in mammals and birds that reflects adaptation to a cold climate, and it is codified as Bergmann's rule. Hokkaido Island is at the southernmost edge of the brown bear‘s distribution in Eurasia, and the bears on this island are smaller in body size and have darker coats than continental bears. In addition, the body size variation can be attributed also to the differences of food habits among areas.

The brown bear shows sexual dimorphism; males are generally much larger than females at the same age and have more extensive territories. Female bears are more conservative, with narrow territories near those of their mothers, whereas males have a wider range of movement, sometimes traveling tens of kilometers a day. The social system is thus one of female philopatry and male dispersal.

Another specific feature of the brown bear is hibernation in underground dens in mountainous areas during winter, generally from December to March. Among mammals, this is the largest species to hibernate. In the wild, bears gorge themselves with food in the autumn to store body fat, but eat nothing during hibernation. Pregnant females generally bear 1–3 small cubs and nurse them in the den. Hibernation is an energy-conservation adaptation that allows survival in a cold environment during winter.

3.2 Phylogeography of the Brown Bear in the Northern Hemisphere

Global Mitochondrial DNA Phylogeny of Brown Bears

The mitochondrial DNA (mtDNA) phylogeography of brown bears in the Northern Hemisphere has been studied for the past 30 years. Because this haploid molecule shows no recombination from one generation to the next, it is a good marker for examining relationships among populations and tracing their migration history.

Given the female philopatry of brown bears, investigating the geographic distribution of maternally inherited mtDNA lineages provides invaluable insights into the migration history of the bears across their distributional range. The present-day distributional patterns of brown bear mtDNA lineages likely reflect the expansion and retreat of populations in response to paleoenvironmental changes. In addition, direct evidence of past distributions can sometimes be obtained from analyses of ancient DNA from bear fossils excavated from caves and permafrost-bearing strata.

Previous phylogenetic studies of brown bear mtDNA have reported at least five lineages (Clades 1–5, Fig. 3.1) (see Hirata et al. 2013). Globally, there are two main mtDNA lineages, Western and Eastern. The Western lineage consists of Clades 1 and 2. Clade 1 in Western Europe comprises two subclades, Clade 1a (Iberian lineage) and Clade 1b (Balkan lineage). Clade 2a comprises brown bears inhabiting the ABC islands in Alaska, North America, while Clade 2b represents the polar bear (*Ursus maritimus*), a distinct species closely related to the brown bear, possibly indicating introgression of mtDNA from brown bears to polar bears during speciation. Clades 3 and 4 are discussed in the following section. Clade 6 is restricted to Middle Eastern and Central Asia, and needs further study phylogeographically.

Triple mtDNA population Structure of Brown bears on Hokkaido

Interestingly, the brown bear population on Hokkaido Island, northern Japan, comprises three distinct mtDNA lineages: Cluster A, equivalent to Clade 3a2; Cluster B, equivalent to Clade 3b; and Cluster C, equivalent to Clade 4 (Fig. 3.2) (Matsuhashi et al. 1999). These lineages are distributed in different areas of Hokkaido, with clear boundaries between them: Cluster A in north-central Hokkaido, Cluster B in eastern Hokkaido, and Cluster C in southern Hokkaido (Fig. 3.3). Brown bears in Europe also comprise three distinct mtDNA lineages, Clades 1a, 1b, and 3a1 (Taberlet and Bouvet 1994). It is noteworthy that the three mtDNA lineages on Hokkaido (83,450 km^2) inhabit an area similar to that of Ireland (84,420 km^2), much smaller than the European continent.

In contrast to lineages in Eurasia and North America, the lineages on Hokkaido also occur in more-distant areas, as shown in Fig. 3.1 (Matsuhashi et al. 2001; Hirata

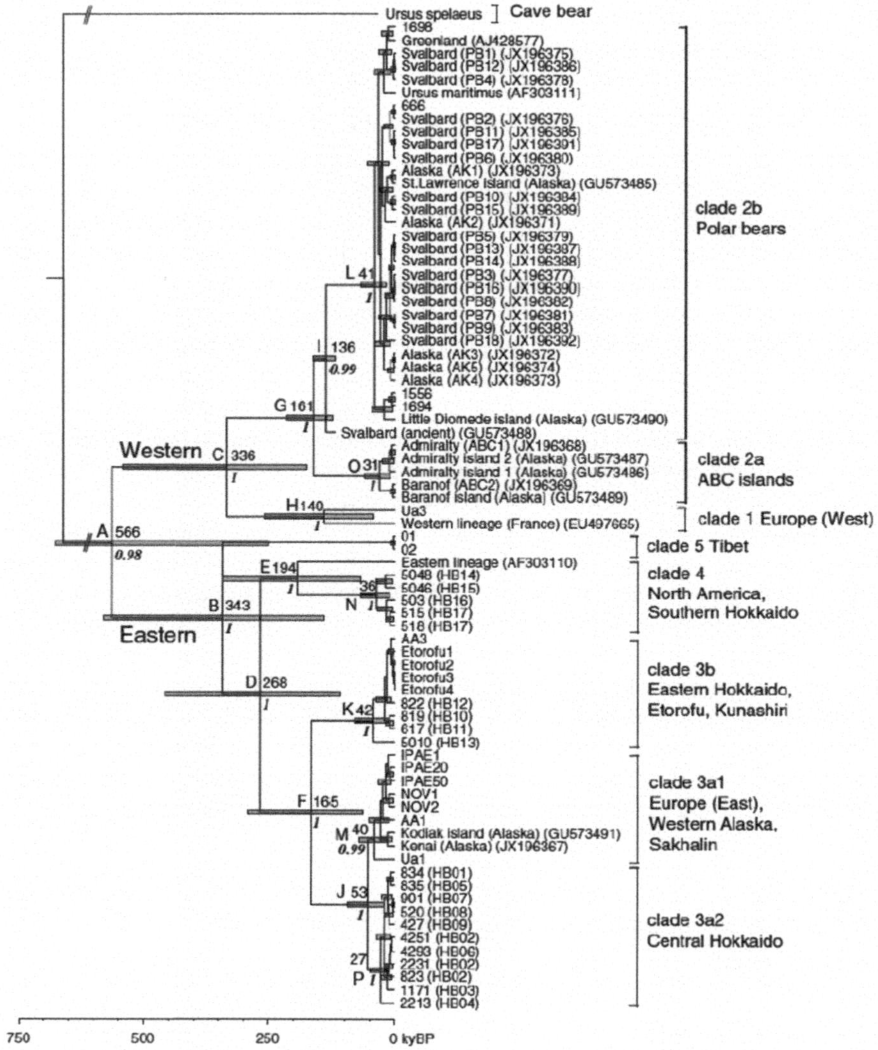

Fig. 3.1 Maximum clade-credibility tree from a BEAST Bayesian analysis of complete mtDNA sequences from brown bears. Major mtDNA clades and their geographic regions are labeled. High posterior-probability support values (0.98–1) are labeled in italics near nodes. The numbers at nodes A–P indicate the estimated mean age in kyBP (1000 of years before present); the horizontal bars across nodes indicate the 95% highest posterior density of nodal age estimates. Terminal taxa are indicated by previously reported sample names, sample numbers, and/or GenBank accession numbers. Haplotype numbers in parentheses for the Hokkaido brown bear individuals correspond to those of Matsuhashi et al. (1999) (also see Fig. 3.2, this chapter). (From Hirata et al. (2013), with permission)

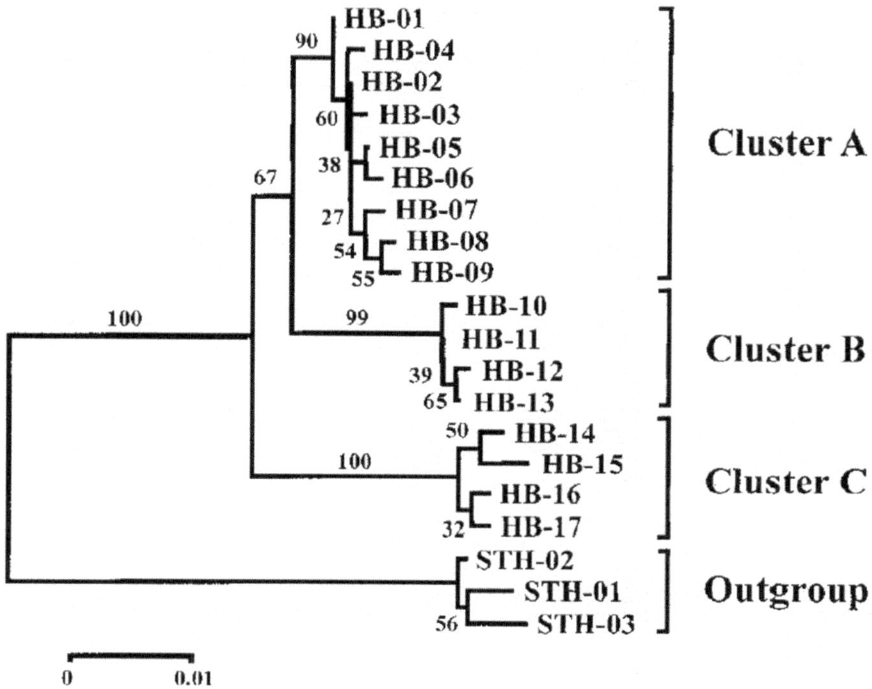

Fig. 3.2 Neighbor-joining tree for 17 mtDNA haplotypes detected in the Hokkaido brown bears (HB-01 to HB-17), with haplotypes from Asiatic black bears (STH-01 to STH-03) as the outgroup. Numbers near nodes are bootstrap values from analyses of 1000 pseudoreplicates. The scale indicates branch length in nucleotide substitutions per site (Kimura two-parameter distances). (From Matsuhashi et al. (1999), with permission)

et al. 2013): Clade 4 (southern Hokkaido) occurs also in the Rocky Mountains, western USA; Clade 3b (eastern Hokkaido) occurs also in eastern Alaska; and Clade 3a2 (north-central Hokkaido) is the sister lineage to Clade 3a1, distributed from eastern Europe across Siberia to western Alaska. The relationships among the mtDNA lineages suggest that the first immigration from continental populations to Hokkaido was from Clade 4 (Cluster C) to southern Hokkaido, followed by immigration from Clade 3b (Cluster B) to eastern Hokkaido, and then Clade 3a2 (Cluster A) to north-central Hokkaido (Fig. 3.3).

An APLP (amplified PCR-fragment length polymorphism) mtDNA analysis revealed the presence of Clade 3b in the inner-continental Altai and Caucasus regions of Eurasia (Hirata et al. 2014). Another APLP analysis of archaeological samples (Mizumachi et al. 2021) showed that the brown bear lineage on Sakhalin Island, north of Hokkaido Island, was Clade 3a1, the same lineage occurring on present-day Sakhalin (Hirata et al. 2013). Although Soya (La Pérouse) Strait between Sakhalin and Hokkaido comprises a geographic boundary between Clades 3a1 and 3a2, Mamiya (Tatar) Strait between Sakhalin and the Eurasian continent is

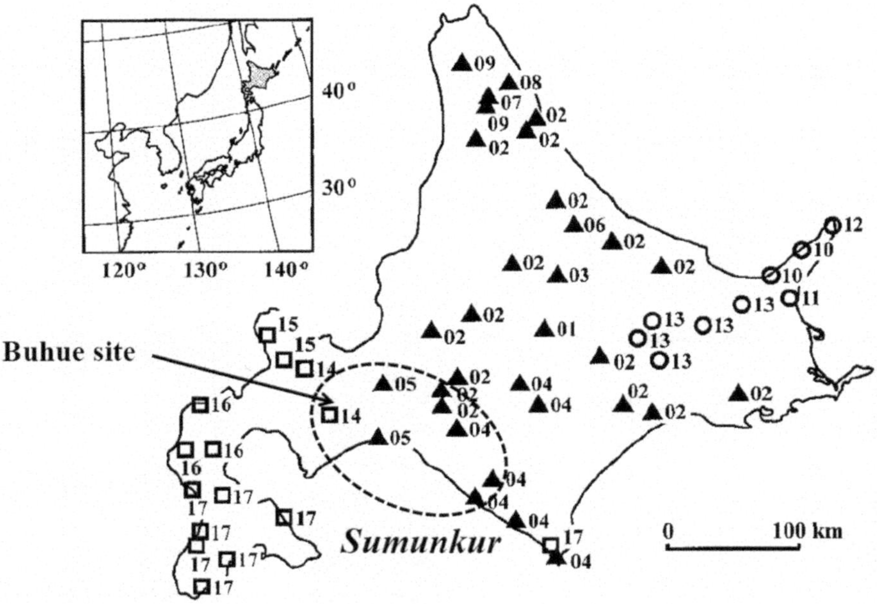

Fig. 3.3 Geographical distribution of mtDNA haplotypes on Hokkaido Island. Triangles, Cluster A; circles, Cluster B; squares, Cluster C. Numbers to the right of the symbols indicate the haplotype number, prefixed by HB in Fig. 3.2. The dashed ellipse shows the area of the *Sumunkur* Ainu communication group; the arrow indicates the location of the Bihue ritual site. (Adapted from Matsuhashi et al. (1999), with permission)

not a geographic boundary for Clade 3a1 (detected both at Sakhalin and at the coastal area of the continental Far East) (Mizumachi et al. 2021; Fig. 3.4).

Recent studies have thus clarified that the Hokkaido brown bears have a complex, dynamic migration history compared to continental populations. Further understanding of the migration history of the Hokkaido bear population can contribute to global evolutionary studies of brown bears and the causal relationships between past climatic changes and bear migrations.

Population Structure of the Hokkaido Brown Bears, Based on Analyses of Y-Chromosomal DNA

Analyses of paternally inherited Y-chromosomal DNA sequences and Y-linked microsatellite genotypes (Hirata et al. 2017) detected a cluster of Y-chromosomal DNA haplotypes in male bears on Hokkaido that was largely exclusive of the haplotypes from continental bears, but no intra-population structure on Hokkaido (Fig. 3.5). This was very different from the maternally inherited mtDNA phylogeography that showed a triple population structure on Hokkaido. This difference might have arisen through male-mediated gene flow among local natal populations, combined with founder effects and subsequent genetic drift after immigration (Hirata et al. 2017).

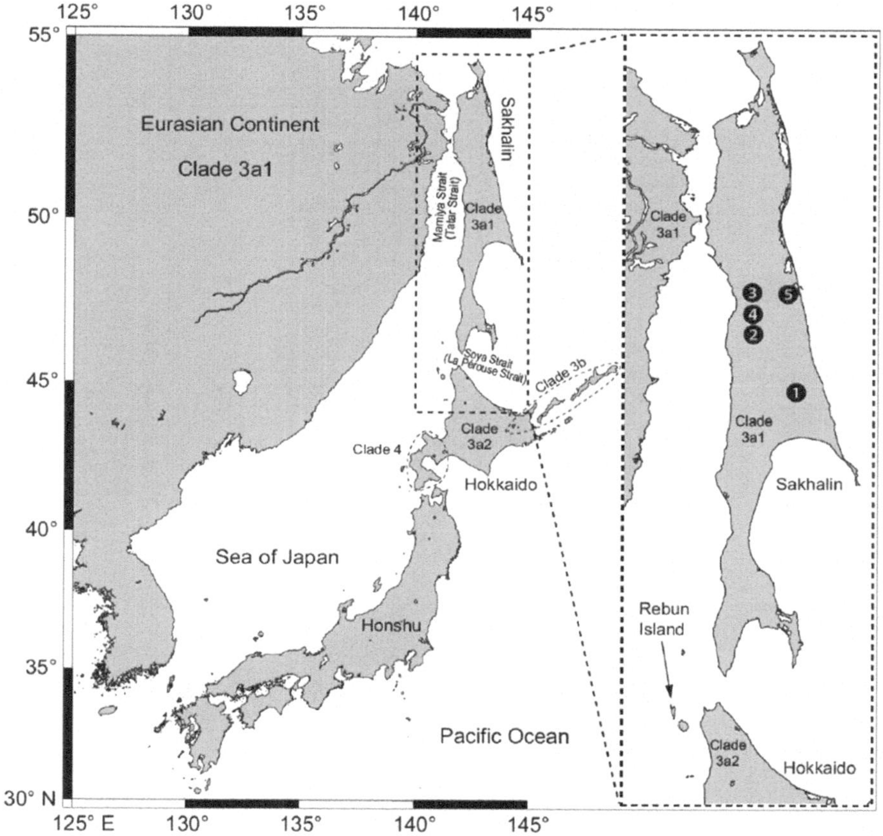

Fig. 3.4 Sampling locations for archaeological and modern brown bears on Sakhalin Island (black circles with the sample identity numbers): (1) Smirnykh region (modern) and Ostantseyava Cave (archaeological), Vaida Mountain; (2) Beloye Village, Tymovsky region (modern); (3) Puzi-1 site, Ado-Tymovo (archaeological); (4) Ado-Tymovo (modern); (5) Ygvo-village, Nogliksky region (modern). The distributions of previously reported mtDNA clades are also indicated on the map. (From Mizumachi et al. (2021))

Population Structure and Demographic History of Hokkaido Brown Bears, Based on Whole-Genomic Analysis

A whole-genomic analysis of Hokkaido brown bears allowed the comparison of single-nucleotide polymorphisms (SNPs) with previously reported genomic data from continental brown bears (Endo et al. 2021). Among six bears analyzed from three areas on Hokkaido, representing the three different mtDNA lineages on the island, all were closely related to one another but genetically distinct from continental brown bears (Fig. 3.6). An analysis of historical demography indicated no increase (relatively stable) in population size for the Hokkaido bear population during the Eemian interglacial period (130,000–114,000 years ago), whereas the continental populations were increasing at this time (Fig. 3.7). In a phylogeny

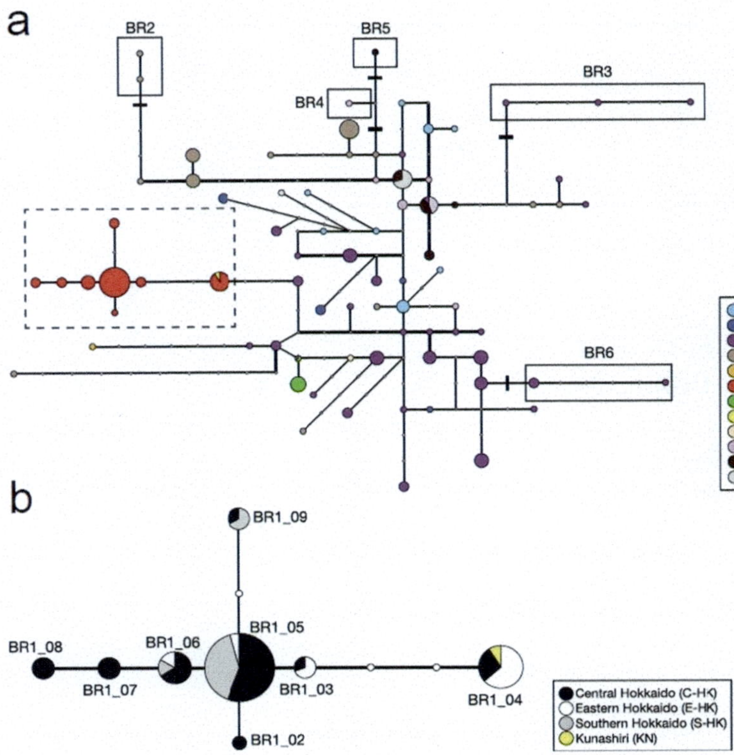

Fig. 3.5 (**a**) Median-joining (MJ) haplotype network for brown bears, based on Y-chromosomal compound haplotypes combined with Y-linked SNPs from a 3.1 kb data set and Y-linked microsatellites. Haplotypes enclosed by the dashed line are from Hokkaido (including one Kunashiri brown bear). Haplotypes enclosed by solid lines denote the same haplotypes (BR2–BR6) as those distinguished by only the 3.1 kb Y-linked data set. The remaining haplotypes have the same haplotype (BR1) distinguished by the 3.1 kb data set. Each color represents a different brown bear population. Each black bar crossing a network line denotes one mutational step revealed by a single nucleotide polymorphism. Small, open circles indicate intermediate haplotypes not actually observed; each line connecting two haplotype circles represents a single microsatellite mutational step. The size of each colored haplotype circle is proportional to the number of individuals having that haplotype. (**b**) MJ network for Hokkaido brown bears, based on the same data as above. (From Hirata et al. (2017))

reconstructed from SNP data, the Hokkaido bears formed a different clade from the European and North American bear populations (Endo et al. 2021). The results suggest that the Hokkaido bears maintained genetic diversity after their geographical isolation on Hokkaido. The genetic uniformity within the Hokkaido population could have resulted from male-mediated gene flow; this result is very different from the phylogenetic relationships reconstructed from mtDNA data (Hirata et al. 2013), but is congruent with the result from the Y-chromosomal DNA data (Hirata et al. 2017) presented in the previous section.

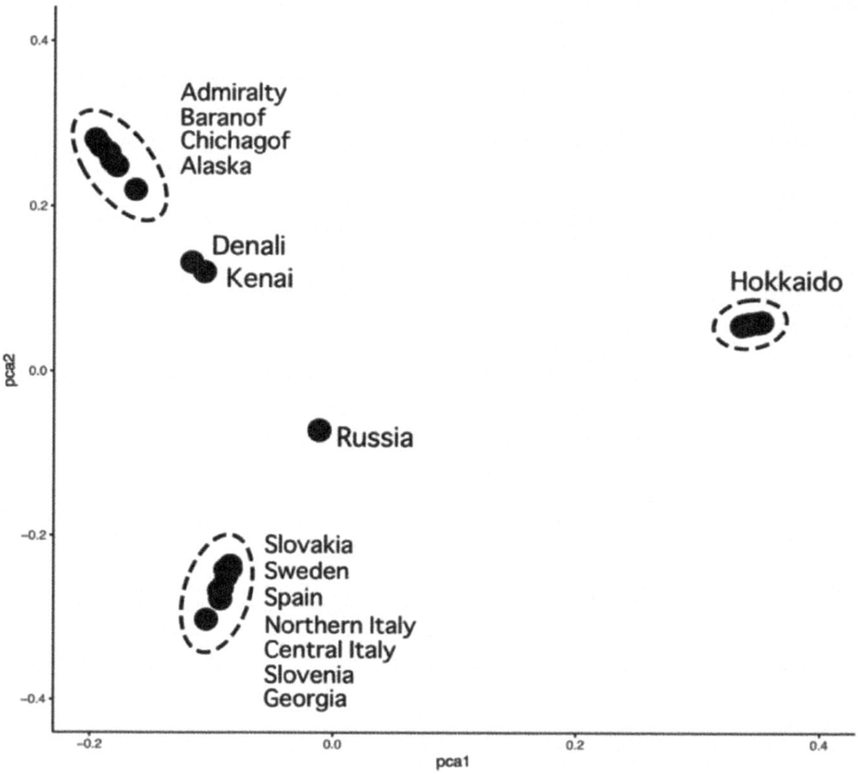

Fig. 3.6 Result of a principal component analysis (PCA) of Hokkaido and continental brown bears. The horizontal axis shows the first principal component, and the vertical axis the second principal component. (From Endo et al. (2021), with permission)

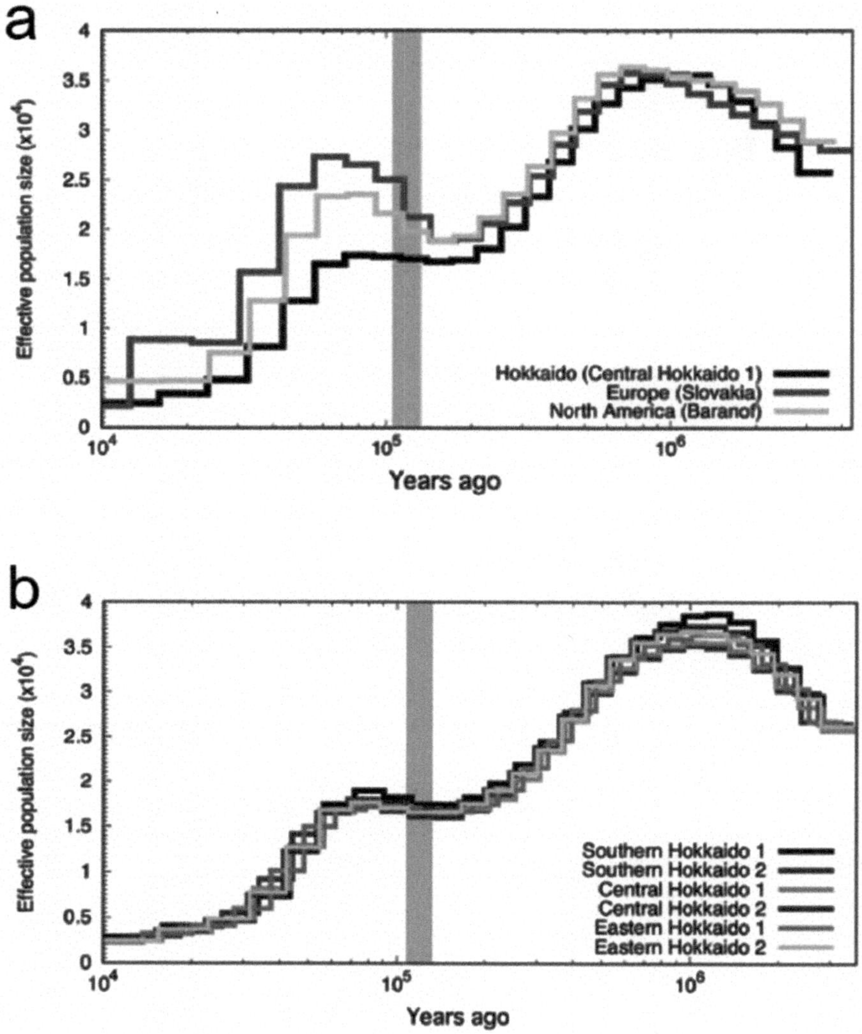

Fig. 3.7 Pairwise sequentially Markovian coalescent (PSMC) estimates of brown bear effective population size (N_e) through time. (**a**) result including bears from central Hokkaido, Europe (Slovakia), and North America (Baranof Island). (**b**) result for six bears from Hokkaido. Vertical gray shading indicates the Eemian interglacial period (13,000–11,400 years ago). (From Endo et al. (2021), with permission)

3.3 Bear-Sending Rituals in the Far East, Including Japan

Two Types of Bear-Sending Rituals

Although the brown bear is a wild animal, undomesticated and rarely captive, it has been widely involved spiritually in human cultures. This bear-human relationship is peculiar to the brown bear and very different from the human relationships to other mammals.

Throughout their history, various native peoples in northern Eurasia and North America have independently developed rituals or ceremonies involving brown bears, across the wide range in the Northern Hemisphere (Hallowell 1926). Initially in their relationship with the brown bear, the native peoples only hunted this species, for food and pelts. It is thus very interesting and important to understand why ceremonialism involving brown bears developed across this wide region. This topic has until now been approached mainly through the discipline of cultural anthropology, and many hypotheses have been proposed.

An overview of the bear-sending rituals that developed in the Ainu Culture (seventeenth century–present) on Hokkaido Island will help elucidate the nature of bear ceremonialism. The bear rituals in the Ainu Culture have been studied mainly through archaeology and ethnology (e.g., Utagawa 1989, 2004; Fitzhugh and Dubreuil 1999; Amano 2003). The Ainu people call the brown bear "the God of Mountains" (*kimun-kamuy* in the Ainu language; henceforth, Ainu words are italicized) and consider it spiritually the highest animal. Between spring and autumn, when brown bears are active, people hunted them and, before taking their pelts and eating their flesh, sent their spirit (or soul) from the human world back to the god world by performing sending-rituals. There were two types of Ainu bear-sending rituals: *opunire* and *iomante* (reviewed by Utagawa 1989). It is known that *iomante* had still been performed by the Showa period, but not so often, and currently it is not done.

Opunire, *Sending Ritual for Bears Hunted in Mountains*

Opunire was performed locally, where and when bears were hunted in mountains. In this ritual, people erected *inaw* (willow branches with the bark partially shaved) around a killed bear, prayed for the bear's spirit, and sent it to the mountains; this praying is called *kamuynomi*. While *opunire* involved bears hunted in mountains, the hunters sometimes took the bear carcasses to their villages (*kotan*) to more politely pray for *kamuynomi* (Utagawa 1989). *Opunire*-type rituals using mountain-hunted bears were widespread across Eurasia and North America (e.g. Hallowell 1926; Rockwell 2003).

Many *opunire* sites are known in mountainous forests on Hokkaido (Chitose Board of Education 1984; Amano 2003; Sato 2006). Several sites around Lake

Shikotsu are located under large, rocky walls of welded tuff, where the skulls and other bones of bears were placed after the rituals. These sites are called *iwaya* or *siratcise* ("ritual rock-shelter sites"). The Chitose Board of Education (1984) conducted an archaeological excavation at the Bihue rock shelter, a relatively large site near the western shore of Lake Shikotsu, and determined the gender, ages, and seasons of death of the bears by examining skull morphology and canine cementum layers. This site was used for *opunire* until about 80 years ago. The bear remains have been preserved by the Chitose Board of Education.

Phylogeographical analyses using mtDNA haplotypes can elucidate the origins of bear remains found in ritual sites, that is, detect from which of the three populations on Hokkaido the bears came, and an mtDNA analysis of skulls from the Bihue ritual site revealed interesting insights into the history of the Ainu community (Masuda et al. 2006). The Bihue site lies within the geographical area of Cluster C (= Clade 4; southern Hokkaido lineage; Matsuhashi et al. 1999; see Fig. 3.3). Among 10 adult skulls (2–24 years old at time of death; killed from March to June) at this site, four were haplotype HB14, characteristic of Cluster C; six were haplotype HB02, HB04, or HB05, characteristic of Cluster A (= Clade 3a2; north-central Hokkaido); but none were from Cluster B (= Clade 3b; eastern Hokkaido) (Masuda et al. 2006). Interestingly, the geographical distribution of these four haplotypes exactly matches the area of *Sumunkur*, an Ainu communication group extending from the Ishikari lowland to Lake Shikotsu and thence to the western part of the Hidaka district (Kaiho 1974) (Fig. 3.3). In other words, the Bihue site was used for *opunire* by people from multiple Ainu villages, promoting intracultural communication and unity via bear-sending rituals (Masuda et al. 2006).

Iomante, *Sending Ritual for Juvenile Bears Nursed at* Kotan

The other type of Ainu bear-sending ritual was *iomante*, which involved juvenile bears (Utagawa 1989). In the spring, just after bears emerge from hibernation in the mountains, it is relatively easy to hunt them because they are undernourished and weak. As noted above, the bears at the Bihue ritual site had mostly been killed in spring (March to June). In addition, during hibernation, pregnant females give birth in their dens and emerge with their cubs in spring. When mother bears were hunted in the mountains, *opunire* was performed on them there. By contract, the juveniles were usually not killed there, but were taken to the *kotan*, where they were reared in wooden cages (*hepereset*) until the autumn of the same or the following year, and then sacrificed at the *kotan* in the *iomante* bear-sending ritual.

Iomante may have derived from *opunire*, which originated directly from hunting in the wild, and the two types of ritual were culturally related. Although *opunire* was performed by a few people when they killed an adult bear in the mountain, people keeping juvenile bears at a *kotan* could decide exactly when they wished to perform *iomante* and then invite the community at large. Since performing the *iomante* ritual at *kotan* much more effectively strengthened community bonds, this type of ritual may have played a leading role in development and establishment of the Ainu Culture. Understanding the origin and history of *iomante* is a dominant theme in Ainu archaeology and ethnology.

Rituals with Captive Juvenile Bears in the Okhotsk Culture, Possibly Leading to the **Iomante** *Ritual*

The Okhotsk Culture (sixth–eleventh centuries CE) was primarily a fishing culture along the southern coastal region of the Okhotsk Sea (northern and eastern Hokkaido; Rebun and Rishiri Islands; southern Sakhalin; southern Kuril Islands). In a zooarchaeological study, Masuda et al. (2001) used genetic data to investigate the origin of a bear-sending ritual that used juvenile bears on Rebun Island, northern Hokkaido.

One of the most famous archaeological localities for the Okhotsk Culture is the Kafukai 1 site on Rebun Island (Fig. 3.4); many faunal remains, including sea animals and brown bears, have been excavated at this site (Oba and Ohyi 1976, 1981). There is no natural population of brown bears on Rebun at present, nor have bear fossil remains yet been found there. The only bear remains on Rebun have come from skull mounds near archaeological house traces, with some of the skulls bearing holes presumably made for ritual purposes. All bear remains at the Kafukai 1 site thus appear to have been brought from outside Rebun Island for use in cultural activities. Although Ohyi et al. (1980) had determined by morphological examination the age at death, season of death, and gender of the excavated bear skulls, their source was unknown.

To clarify the source of bear remains in the Kafukai 1 site, Masuda et al. (2001) conducted a zooarchaeological study to compare ancient mtDNA haplotypes with modern haplotype data from Hokkaido Island, of which distribution patterns were thought to have been established prior the Okhotsk Cultural Period, based on brown bears' migration history to Hokkaido as already mentioned. The ancient brown bears came from two mtDNA haplotype lineages: Cluster A (Clade 3a2; north-central Hokkaido) and Cluster C (Clade 4; southern Hokkaido) (see Figs. 3.2, 3.3, and 3.4). Haplotypes of Cluster A were identified mainly from adults more than 2 years old, sacrificed in the spring. This seemed reasonable, because Cluster A haplotypes are distributed in northern Hokkaido, which was also inhabited by the Okhotsk Culture. The ancient people could have hunted adult bears on Hokkaido and taken them (probably as body parts) to Rebun Island for rituals.

By contrast, a Cluster C haplotype (HB14; southern Hokkaido) was found in juveniles less than 1 year old, sacrificed in autumn. This indicated that people hunted mother-cub pairs in southern Hokkaido, but that only cubs were transported to Rebun. Southern Hokkaido was inhabited by the Epi-Jomon/Satsumon Culture, which was distinct from the Okhotsk Culture but instead had affinities to the Tohoku district on northern Honshu Island. The people of the Okhotsk Culture on Rebun

Island probably imported and reared juvenile bears until the performance of rituals in autumn. For this, people of the Okhotsk Culture may have visited southern Hokkaido to receive juvenile bears from those of southern Hokkaido. People in different ancient cultures thus seem to have similarly imparted significance to and placed high value on juvenile bears, which functioned in intercultural communication and bonding (Masuda et al. 2001).

In a study of prehistoric bears on Sakhalin, Mizumachi et al. (2021) showed that the bear remains at the Kafukai 1 site did not come from Sakhalin, but surely came from Hokkaido Island. Thus, multidisciplinary genetic and archaeological studies of the Kafukai bear remains suggest that the origin of *iomante* rituals (using reared juvenile bears) can be traced to the Okhotsk Culture. Further archaeological studies on the history of the different cultures and modes of communication between them are necessary.

Acknowledgements I thank Dr. Matthew H. Dick (Hokkaido University) for commenting on and editing the manuscript for this chapter.

Competing Interests The author has no relevant financial or non-financial interests to disclose.

References

Amano T (2003) The origin of the bear ceremony. Yuzankaku, Tokyo. (in Japanese)

Chitose Board of Education (1984) Archaeological researches in Bihue, Chitose City: Chitose archaeological record 10. Chitose Board of Education, Hokkaido. (in Japanese)

Endo Y, Osada N, Mano T, Masuda R (2021) Demographic history of the brown bear (*Ursus artcos*) on Hokkaido Island, Japan, based on whole-genomic sequence analysis. Genome Biol Evol 13(9):evab195. https://doi.org/10.1093/gbe/evab195

Fitzhugh WW, Dubreuil CO (eds) (1999) Ainu: Spirit of a northern people. Washington, DC and Seattle, WA, National Museum of Natural History, Smithsonian Institution, and University of Washington Press

Hallowell AI (1926) Bear ceremonialism in the northern hemisphere. Am Anthropol 28:1–175

Hirata D, Mano T, Abramov AV, Baryshnikov GF, Kosintsev PA, Vorobiev AA, Raichev EG, Tsunoda H, Kaneko Y, Murata K, Fukui D, Masuda R (2013) Molecular phylogeography of the brown bear (*Ursus arctos*) in northeastern Asia based on analyses of complete mitochondrial DNA sequences. Mol Biol Evol 30:1644–1652. https://doi.org/10.1093/molbev/mst077

Hirata D, Abramov AV, Baryshnikov GF, Masuda R (2014) Mitochondrial DNA haplogrouping of the brown bear, *Ursus arctos* (Carnivora: Ursidae) in Asia, based on a newly developed APLP analysis. Biol J Linn Soc 111:627–635. https://doi.org/10.1111/bij.12219

Hirata D, Mano T, Abramov AV, Baryshnikov GF, Kosintsev PA, Murata K, Masuda R (2017) Paternal phylogeographic structure of the brown bear (*Ursus arctos*) in northeastern Asia and the effect of male-mediated gene flow to insular populations. Zool Lett 3:21. https://doi.org/10.1186/s40851-017-0084-5

Kaiho M (1974) The logic behind history of northern Japan. Yuzankaku, Tokyo. (in Japanese)

Masuda R, Amano T, Ono H (2001) Ancient DNA analysis of brown bear (*Ursus arctos*) remains from the archeological site of Rebun Island, Hokkaido, Japan. Zoolog Sci 18:741–751. https://doi.org/10.2108/zsj.18.741

Masuda R, Tamura T, Takahashi O (2006) Ancient DNA analysis of brown bear skulls from a ritual rock shelter site of the Ainu culture at Bihue, Central Hokkaido, Japan. Anthropol Sci 114:211–215. https://doi.org/10.1537/ase.051219

Matsuhashi T, Masuda R, Mano T, Yoshida MC (1999) Microevolution of the mitochondrial DNA control region in the Japanese brown bear (*Ursus arctos*) population. Mol Biol Evol 16:676–684. https://doi.org/10.1093/oxfordjournals.molbev.a026150

Matsuhashi T, Masuda R, Mano T, Murata K, Aiurzaniin A (2001) Phylogenetic relationships among worldwide populations of the brown bear *Ursus arctos*. Zoolog Sci 18:1137–1143. https://doi.org/10.2108/zsj.18.1137

Mizumachi K, Gorbunov SV, Vasilevski AA, Amano T, Ono H, Kosintsev PA, Hirata D, Nishita Y, Masuda R (2021) Phylogenetic relationships of ancient brown bears (*Ursus arctos*) on Sakhalin Island, revealed by APLP and PCR-direct sequencing analyses of mitochondrial DNA. Mamm Res 66:95–102. https://doi.org/10.1007/s13364-020-00542-7

Oba T, Ohyi H (1976) Research of the Okhotsk culture 2: archeological site of Kafukai, part 1. University of Tokyo Press, Tokyo. (in Japanese)

Oba T, Ohyi H (1981) Research of the Okhotsk Culure 3: archeological site of Kafukai, part 2. University of Tokyo Press, Tokyo. (in Japanese with an English abstract)

Ohyi H, Ohtaishi N, Nishimoto T (1980) Determination of age, sex and mortal-season of Yezo brown bear, excavated from the Kabukai a site, Rebun Island, Hokkaido. Res North Cult (Hoppou Bunka Kenkyu) 13:43–74. (in Japanese with an English abstract)

Rockwell D (2003) Giving voice to bear: north American Indian myths, rituals, and images of the bear (paperback edition). Roberts Rinehart Publishers, Lanham, MD

Sato T (ed) (2006) Ethno-archaeological studies on the rock shelters used by the Ainu: researches on the bear-hunting and the sending ceremony performed in the upper reaches of the Izari River basin, Hokkaido. Rokuitsushobo, Tokyo. (in Japanese)

Taberlet P, Bouvet J (1994) Mitochondrial DNA polymorphism, phylogeography, and conservation genetics of the brown bear *Ursus arctos* in Europe. Proc R Soc B 255:195–200. https://doi.org/10.1098/rspb.1994.0028

Utagawa H (1989) Archaeology of Iomante. University of Tokyo Press, Tokyo. (in Japanese)

Utagawa H (ed) (2004) Iomante of bears and owls—Ethnoarchaeology of Ainu. Doseisha, Tokyo. (in Japanese)

Chapter 4
The Japanese Wolf and Dogs

Yohey Terai

Abstract In this chapter, I introduce the relationship between the Japanese wolf (*Canis lupus hodophilax*, (Temminck, Tijdschr Natuurl Gesch Physiol 5:273–293, 1839)) and the dog, revealed by the whole genome information. The Japanese wolf is a lineage of the Asian gray wolf that once inhabited the southern part of the Japanese archipelago (Honshu, Shikoku, and Kyushu) until about 100 years ago when it became extinct. A recent study showed that the Japanese wolf lineage diverged from other gray wolves in East Asia around 17,000–40,000 years ago. The Japanese wolf is the most closely related lineage to the dog, and the Japanese wolf ancestor is estimated to have interbred with the ancestor of the East Eurasian dog. Therefore, the genome of the East Eurasian dog contains the genome of the ancestor of the Japanese wolf. The genome analysis of the museum specimen showed that a hybrid between the dog and Japanese wolf existed in the Edo period, suggesting that the dog and Japanese wolf may have had a close relationship in the Japanese archipelago. Although there are still many mysteries about the Japanese wolf, we expect that ancient genome data using dog/wolf archaeological remains from the Japanese archipelago will further clarify the relationship between the Japanese wolf and the dog.

Keywords Dog · Japanese wolf · Whole genome analysis · Ancient DNA · Introgression

Abbreviation

NGSD New Guinea singing dog

Y. Terai (✉)
SOKENDAI (The Graduate University for Advanced Studies), Research Center for Integrative Evolutionary Science, Hayama, Kanagawa, Japan
e-mail: terai_yohei@soken.ac.jp

N. Osada et al. (eds.), *Phylogeographic History of Plants and Animals Coexisting with Humans in Asia*, Evolutionary Studies,
https://doi.org/10.1007/978-981-97-6887-5_4

4.1 Introduction

The Japanese wolf (*Canis lupus hodophilax*, Temminck 1839) is a subspecies of the gray wolf that inhabited Honshu, Shikoku, and Kyushu in Japan. The gray wolf (*C. lupus*) includes the North American and Eurasian populations, the Japanese wolf, and the dog. The Japanese wolf went extinct about 100 years ago, and there are only three taxidermied specimens in Japan and one in the Netherlands, so there are few clues left to learn about them. Philipp Franz von Siebold and his assistants, who stayed at Dejima in Nagasaki Prefecture during the Edo period, sent the Japanese wolves to the Netherlands. Siebold's three specimens of animals, which were called dog, wolf, and yamainu, respectively (Ishiguro et al. 2021). These three individuals were described as *C. lupus hodophilax* by Coenraad Jacob Temminck (1839): note, the original description is *Canis hodophilax*, but it is given here as *Canis lupus hodophilax* to avoid confusion. These three specimens are kept at the Naturalis Biodiversity Center in the Netherlands. In Hokkaido, in the northern part of the Japanese archipelago, a species of wolf called the Ezo wolf (*C. l. hattai* Kishida 1931) inhabited the area, but this species also went extinct more than a 100 years ago (Inukai 1982). Mitochondrial DNA analysis has shown that the Ezo wolf is genetically closer to the dog/gray wolf than the Japanese wolf (Matsumura et al. 2014). Mitochondrial DNA (Segawa et al. 2022) and nuclear genome (Gojobori et al. 2024) analyses indicate that the giant wolves that inhabited Japan more than 10,000 years ago are of a different lineage from the Japanese wolf.

As explained above, there are few clues to the extinct Japanese wolf. However, genetic information in DNA can still be available today. DNA is a highly stable molecule and has been analyzed from bones that are thousands of years old (Willerslev and Cooper 2005). Even 1 million years old material can be analyzed if it is well preserved, such as in permafrost (van der Valk et al. 2021). DNA has been extracted from Japanese wolf bones and other specimens, and mitochondrial DNA has been analyzed (Ishiguro et al. 2009, 2016; Matsumura et al. 2014, 2021). Recently, a study reported by Gojobori et al. (2024) used DNA from these mitochondrial DNA studies and determined the whole nuclear genome sequence of nine Japanese wolves. This review focuses on the content of the whole genome analysis of the Japanese wolf and introduces the relationship between the Japanese wolf and dogs.

4.2 The Japanese Wolf as a Subspecies of the Gray Wolf

The gray wolf is a widely distributed species from North America to Eurasia (Loog et al. 2020). The Japanese wolf is a subspecies of the gray wolf. To learn about the Japanese wolf, I begin to explain the evolution of the gray wolf. Gray wolves form a monophyletic group, suggesting they evolved from a single ancestor (Fan et al. 2016; Leathlobhair et al. 2018). The most closely related outgroup to the gray wolf has been reported to be the golden jackal (*Canis aureus*) based on mitochondrial DNA analysis (Koepfli et al. 2015) and the North American coyote (*Canis latrans*) based on nuclear genome analysis (Gojobori et al. 2024; Perri et al. 2021). Although results often differ between phylogenetic analyses of mitochondrial DNA and nuclear genomes, considering the amount of information available, the coyote is the most closely related outgroup (Fig. 4.1). The origin of the gray wolf is likely to be North America since its most closely related outgroup, coyote, inhabits in North America. Dire wolves (*Aenocyon dirus*), an animal morphologically similar to the gray wolf, had inhabited North America. Recently, DNA was extracted from the bones of dire wolves, and the whole genome sequence (ancient genome) was analyzed (Perri et al. 2021). Analysis of the dire wolf genome reported that dire wolves are not closely related to the gray wolf and did not interbreed with coyotes or gray wolves in North America (Perri et al. 2021).

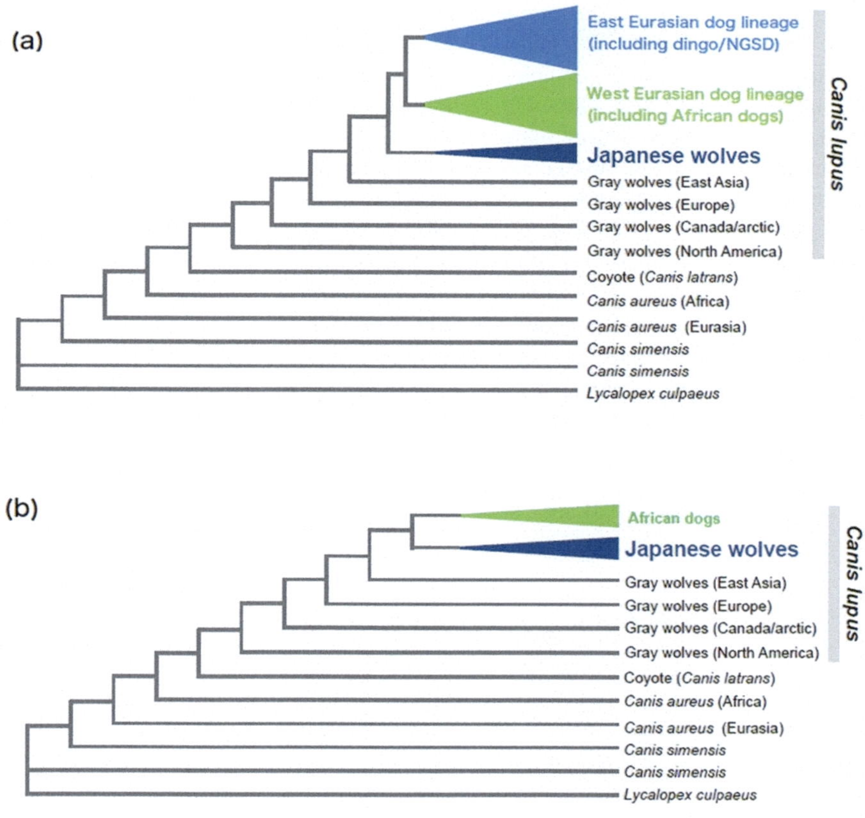

Fig. 4.1 Phylogenetic relationships of wolves and dogs based on the whole genome of gray wolves, Japanese wolves, and (**a**) all dogs and (**b**) African dogs. (Modified from Gojobori et al. 2024)

Next, I explain the internal population relationships of the gray wolf. Whole genome studies indicate that a North American lineage diverges first, followed by a lineage including Canadian, and Arctic wolves (Fan et al. 2016; Botigué et al. 2017; Leathlobhair et al. 2018; Gojobori et al. 2024) (see Fig. 4.1a). Considering the divergence of the North American populations (North America and Canada/arctic) at the base of the monophyletic group of the gray wolf and the coyote being the most closely related outgroup of the gray wolf, the origin of the gray wolf is assumed to be North American. The populations (North America and Canada/arctic) diverge at the base of the gray wolf, thus the remaining lineage is the Eurasian lineage. It would be easy to understand if the Eurasian gray wolf population could be explained by phylogenetic relationships alone, but the story is a little more complicated.

Since the Eurasian gray wolf lineage includes a dog lineage, individuals from different lineages would have encountered each other on the Eurasian continent. Therefore, hybridization would have occurred between different wolf populations, including the dog lineage. The hybrids would then be backcrossed with the original population, resulting in a genetic exchange (gene flow) between the two populations. Two populations with a gene flow become genetically closer. In constructing a phylogenetic tree, the original phylogenetic relationship is affected by a gene flow, leading to an incorrect phylogenetic relationship. Therefore, when examining the phylogenetic relationships of populations with gene flow, it is first necessary to determine which populations have genetic exchange among them. Indeed, European, Middle Eastern, Central Asian, Siberian, and East Asian gray wolf populations have been reported to have had gene flows with the dog populations (Savolainen et al. 2002; von Holdt et al. 2010; Thalmann et al. 2013; Freedman et al. 2014; Shannon et al. 2015; Wang et al. 2016; Sinding et al. 2020). In animals with such a complex history of gene flows, phylogenetic relationships based only on genetic distance cannot distinguish between actual branching and gene flow history. Gojobori et al. (2024) examined the history of gene flows between the Japanese wolf, Eurasian gray wolf populations, and the dogs and showed the phylogenetic relationships without the effects of gene flows. Therefore, before explaining the phylogenetic position of the Japanese wolf, I introduce the history of gene flows among the Eurasian gray wolf populations.

4.3 Gene Flows Among Eurasian Gray Wolf Populations

As mentioned above, to understand the phylogenetic position of the Japanese wolf, the first step should be to clarify the gene flows between the Japanese wolf and other gray wolf populations, including dogs. Principal component analysis using whole genome data from a large number of individuals from the dog, Japanese wolf, and gray wolf populations showed that the Japanese wolf is a genetically unique population (Fig. 4.2). Genomic data analysis of the British Museum's Japanese wolf (reported as the Honshu wolf) initially suggested that the Honshu wolf may be closely related to the Pleistocene wolves in Eurasia (Niemann et al. 2021). However, analysis of the Honshu wolf and seven Japanese wolf genomes revealed that the Honshu wolf belongs to the Japanese wolf population (Gojobori et al. 2024). Analyses of the Japanese wolf and other gray wolf populations have shown that the Japanese wolf had no gene flow not only with geographically isolated populations in North America but also with Eurasian gray wolf populations, including those in East Asia (Gojobori et al. 2024). On the other hand, the Japanese wolf had a gene flow with a dog lineage.

Dogs are roughly divided genetically into the East and West Eurasian lineages (Frantz et al. 2016; Gojobori et al. 2024) (Fig. 4.1a). The West Eurasian lineage includes European and African dogs, while the East Eurasian lineage includes Japanese dogs, Southeast Asian village dogs, Australian dingoes, and New Guinea singing dogs (NGSD) that survived in the highlands of Papua New Guinea (Frantz et al. 2016; Gojobori et al. 2024). Although dogs of the East and West Eurasian lineages have experienced admixture after splitting into two lineages, African dogs in the West Eurasian lineage and dingo and NGSD in the East Eurasian lineage have remained close to the pure dogs of their respective lineages by less mixing (Larson

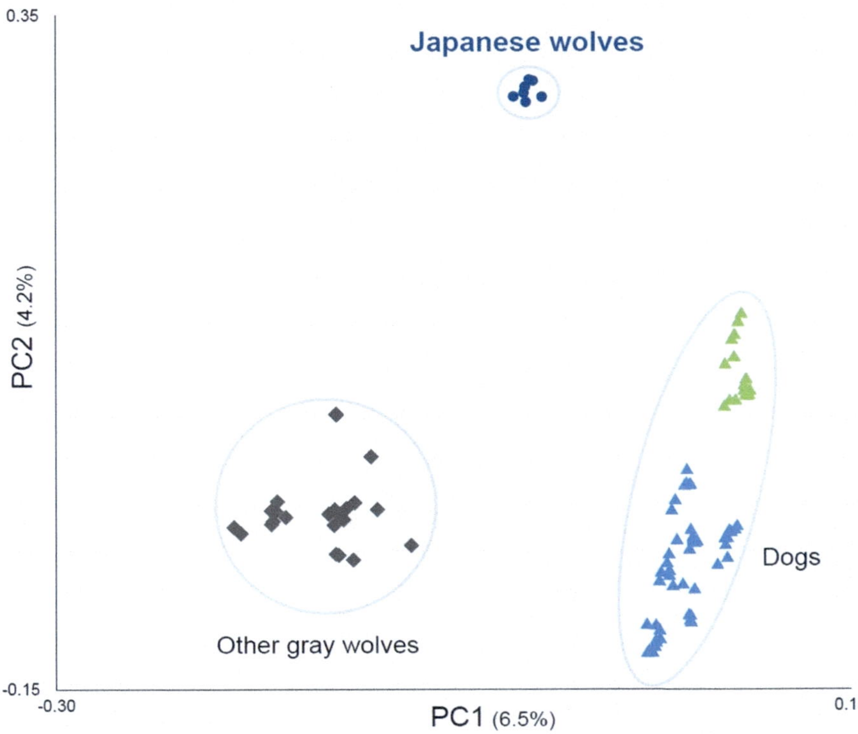

Fig. 4.2 Principal component analysis based on the whole genome data. (Modified from Gojobori et al. 2024)

et al. 2012; Fan et al. 2016; Gojobori et al. 2024). These dogs were geographically distant from the Eurasian continent and thus may have survived without mixing.

Dogs of the East Eurasian lineage, especially with dingo and NGSD, had gene flows with Japanese wolves, but dogs of the West Eurasian lineage had little or no gene flow with Japanese wolves (Gojobori et al. 2024) (Fig. 4.3). However, this result is mysterious when considered carefully, and raises the question of why gene flow was found between the Japanese wolf, which was only found in Japan, and the dogs of South Asia and Australia. Did the dogs once arrive in Japan and then return to the continent, migrating to the highlands of New Guinea and Australia? To answer these questions, it is necessary first to know the phylogenetic relationships among the gray wolves, including the Japanese wolves, and the answers will be explained in the latter part of this chapter.

Gene flows have been shown between Middle Eastern gray wolves and dogs of the West Eurasian lineage (von Holdt et al. 2010; Bergström et al. 2020, 2022; Gojobori et al. 2024). In addition, to a lesser degree than in the Middle Eastern gray wolves, European and Indian gray wolves also show gene flows with dogs of the West Eurasian lineage (Gojobori et al. 2024). In summary, the main gene flow

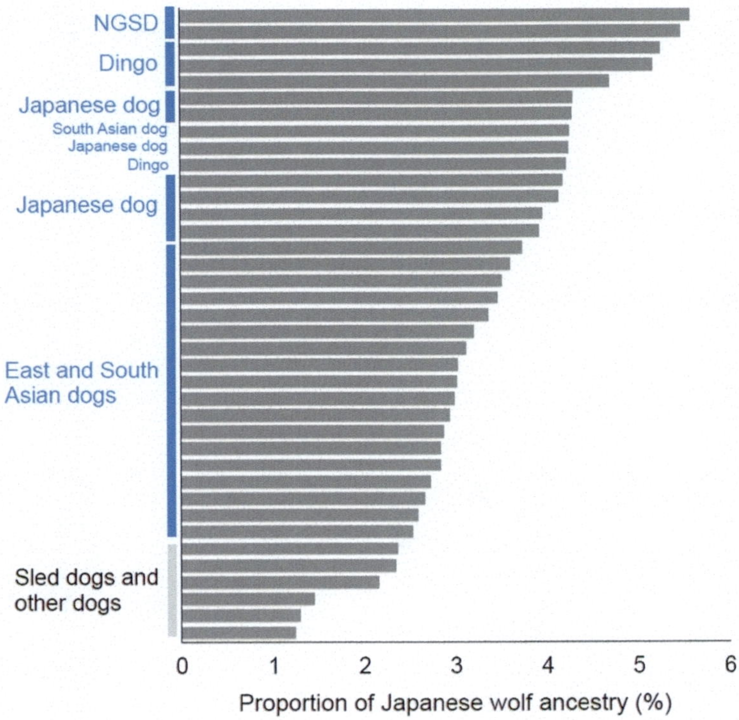

Fig. 4.3 Proportion of the Japanese wolf ancestry in the dog genomes. (Modified from Gojobori et al. 2024)

events were between Japanese wolves and dogs of the East Eurasian lineage and between Middle Eastern gray wolves and dogs of the West Eurasian lineage.

4.4 Phylogenetic Position of the Japanese Wolf

As mentioned previously, in animals with a history of hybridization and gene flow, it is not easy to distinguish between actual branching and the effect of gene flow. Therefore, individuals or populations that may have hybridized must be excluded from the phylogenetic analysis to obtain a correct phylogenetic relationship. To obtain a correct phylogenetic relationship between the Japanese wolf and the Eurasian gray wolves, including dogs, it is best to construct a phylogenetic tree excluding the East Eurasian dogs, which have a gene flow with the Japanese wolf, and the wolves of the Middle East, which have gene flows with the West Eurasian dogs. In particular, it is best to use only African dogs that have not shown any gene flow with the Japanese wolf for the dog lineage. The analysis of the selected populations described above revealed the phylogenetic relationships of the Eurasian gray wolves (Gojobori et al. 2024). In the following, I explain the relationship.

Among the Eurasian gray wolves, the European population diverges at the basal position in the phylogenetic tree (Fig. 4.1b). The second diverging at the basal is the East Asian populations, which includes the Himalayan and Chinese populations. The Japanese wolf/dog monophyletic group forms a sister group to this East Asian populations. Although this phylogenetic tree does not include the Middle Eastern gray wolf and the East Eurasian dogs, the phylogenetic analysis including them shows that the Middle Eastern gray wolf forms a monophyletic clade with the European population, and the East Eurasian dogs form a monophyletic clade with the West Eurasian dogs (Fig. 4.1a).

These phylogenetic trees reveal the phylogenetic position of the Japanese wolf. Japanese wolves form a monophyletic group, independent from the East Asian gray wolf populations and most closely related to the dog monophyletic group. Before showing that Japanese wolves were closely related to dogs, dogs were thought to be derived from the already extinct Eurasian unknown gray wolf population (Larson et al. 2012; Thalmann et al. 2013; Freedman et al. 2014; Skoglund et al. 2015; Fan et al. 2016; Frantz et al. 2016). The phylogenetic trees in Fig. 4.1 show that the Japanese wolf is most closely related to the dog lineage compared to other gray wolf populations, indicating that the dog ancestor diverged from a common ancestor with the Japanese wolf (Gojobori et al. 2024). In addition, the phylogenetic tree shows that the European population is the first divergence in Eurasia, followed by the East Asian population and the most Eastern Japanese wolf (Fig. 4.4). Based on this phylogenetic relationship, the dog lineage diverged between the East Asian gray wolf populations and the Japanese wolf in East Asia (Gojobori et al. 2024). Therefore, we can assume that the common ancestor of the dog and the Japanese wolf inhabited East Asia. After the divergence of the two respective lineages, one lineage was domesticated by humans and became the dog. In contrast, the other lineage became extinct on the continent and survived in the Japanese archipelago, i.e., the Japanese wolf. Although this is still a speculative consideration, if the bones from individuals close to a common ancestor of the dog and Japanese wolf are excavated, ancient DNA analysis will reveal more details about where the dog lineage diverged.

4.5 Contribution of the Japanese Wolf Genome to Dogs in Eastern Eurasia

Returning to the previous section, I discuss the mystery of the gene flow between the Japanese wolf and the East Eurasian dog. Analysis of ancient genomic data has revealed that the 9500-year-old dog genome from Siberia already contained the Japanese wolf genome (Gojobori et al. 2024). In other words, hybridization between the Japanese wolf and the dog occurred at least before 9500 years ago. The earliest dog bones in Japan date back to 9600 years ago (Shigehara and Hongo 2000), suggesting that dogs arrived in the Japanese archipelago with humans earlier than the earliest bones. Analysis of mitochondrial DNA from ancient Japanese dogs also estimates the arrival of dogs to the Japanese archipelago 11,500–9600 years ago (Xiaokaiti et al. 2023). Based on these findings, we can assume that a hybridization event between the ancestor of the Japanese wolf and the dog occurred more than 10,000 years ago in East Asia (not in the Japanese archipelago), and the dog, which

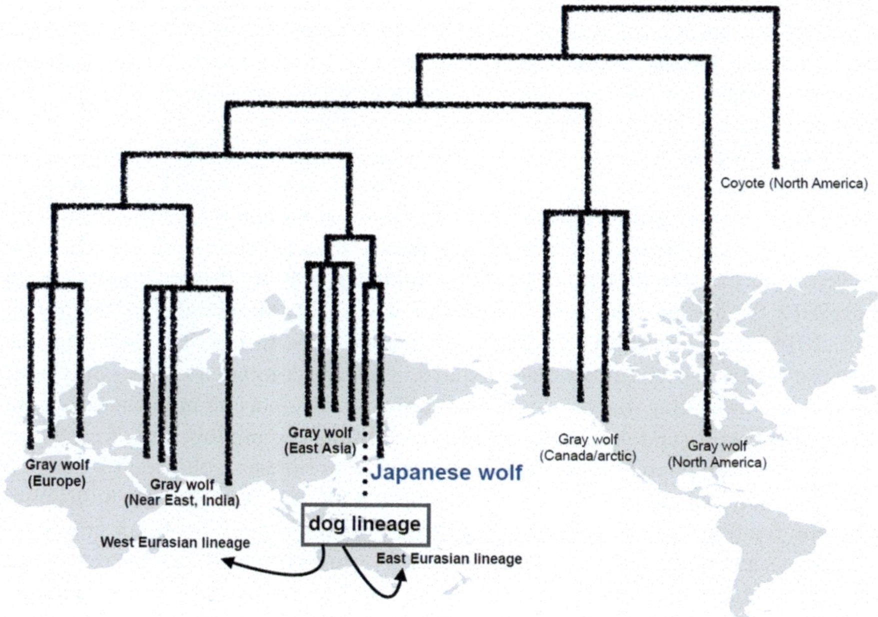

Fig. 4.4 Phylogenetic relationships of wolves and dogs on a world map

already contained the Japanese wolf genome in its genome, migrated to the Japanese archipelago.

Regarding the direction of genomic introgression, Gojobori et al. (2024) showed that the Japanese wolf genome has introgressed into the dog genomes in Eastern Eurasia. The dingo and NGSD contain the highest proportion of the Japanese wolf genome, with a maximum of about 5.5% (Fig. 4.3). Next to them are Japanese dogs such as Akita, Kishu, and Shiba, and native dogs from East and Southeast Asia (Fig. 4.3). On the other hand, no Japanese wolf genome introgression has been detected in European or African dogs (Gojobori et al. 2024). The different proportions of the genome of the Japanese wolf ancestor in the genome of each dog breed raise the question of whether the number of introgression event was one or multiple. The answer was estimated by Gojobori et al. (2024). When examining the genetic affinities of the world's dog breeds to the dingo/NGSD and Japanese wolf, respectively, dogs genetically close to dingo/NGSD are also close to Japanese wolf, and dogs distant from dingo/NGSD tend to be distant from Japanese wolf (Gojobori et al. 2024). These affinity patterns suggest that one genomic introgression event has occurred from the Japanese wolf ancestry to the ancestors of the East Eurasian dogs (see detailed explanation in Gojobori et al. 2024). Taken together, we can estimate the evolutionary process of the Japanese wolf, the dog, and the other gray wolf populations, as shown in Fig. 4.5.

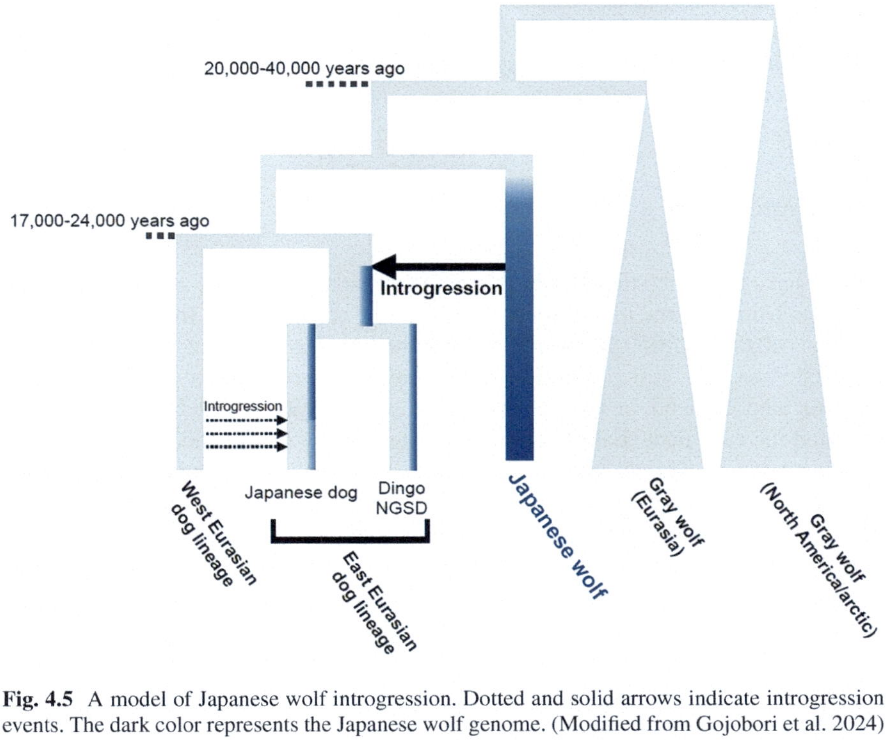

Fig. 4.5 A model of Japanese wolf introgression. Dotted and solid arrows indicate introgression events. The dark color represents the Japanese wolf genome. (Modified from Gojobori et al. 2024)

Next, I explain the estimation of the date when the ancestors of the Japanese wolf inhabited the Eurasian continent. The divergence of the Eurasian gray wolf population (excluding the Japanese wolf) and the dog lineage is estimated to be 20,000–40,000 years ago (Skoglund et al. 2015; Botigué et al. 2017). Next, the divergence of the East-West Eurasian lineage of dogs is estimated to be 17,000–24,000 years ago (Botigué et al. 2017). Since the divergence of the Japanese wolf is after the divergence from the Eurasian gray wolf population and before the divergence of the East-West Eurasian lineage of dogs, the divergence of the Japanese wolf can be estimated to be between 40,000 and 17,000 years ago. After the East-West Eurasian dog lineages diverged, the Japanese wolf introgressed with the ancestor of the East Eurasian lineage of dogs.

Evidence of hybridization between Japanese wolves and dogs has been reported. As mentioned in the introduction, Siebold sent a specimen called Yamainu to the Netherlands during the Edo period. A detailed analyses of the DNA from this individual revealed that the genome contained the genome of the Japanese wolf and the dog (Gojobori et al. 2024). The modern Shiba was the closest to the dog that provided the genome to the Yamainu, with 39% of the genome contained in the Yamainu (Gojobori et al. 2024). If the Yamainu genome can be compared to the genome of a Japanese dog that lived in the Edo period, the proportion of the dog genome in the Yamainu individual might be much higher, perhaps as high as 50%. If it is 50%, the Yamainu is considered an individual of the first-generation hybrid between a dog and a Japanese wolf. Mitochondrial DNA from this Yamainu individual has been reported to be Japanese wolf-type (Matsumura et al. 2021). Since mitochondrial DNA is maternally inherited, we can assume that the mother of the Yamainu individual was a Japanese wolf and the father was a dog.

4.6 Conclusion

So far, I have explained the relationship between Japanese wolves and dogs that can be known from genome information. The Japanese wolf diverged from the Eurasian gray wolf 40,000–17,000 years ago and was one lineage of the gray wolf that inhabited East Asia. The last places where the Japanese wolf survived were Honshu, Shikoku, and Kyushu in the Japanese archipelago. When the Japanese wolves inhabited the continent (East Asia), they introgressed with the ancestor of the East Eurasian lineage of dogs, and the Japanese wolf ancestry genome is still widely present in the genomes of dogs in East Eurasia (including Australia and New Guinea). Since the Japanese wolf is the most closely related lineage to the dog and the Japanese wolf genome is present in the dog genome of Eastern Eurasia, the Japanese wolf has likely played a role in the dog's domestication. Surprisingly, such a unique lineage of gray wolves survived until only a 100 years ago. The isolation of the Japanese archipelago from the continent was probably the reason for their survival. Knowing the importance of the Japanese wolf, it is a loss that they are already extinct.

Acknowledgments First of all, I would like to express my deepest gratitude to Dr. Naotaka Ishiguro for making a deep connection between me and the extinct Japanese wolf, for telling me so much about the Japanese wolf, and for providing the DNA samples of the Japanese wolf for an article (Gojoborid et al. 2024). I would also like to thank Dr. Jun Gojobori, Dr. Nami Arakawa, Dr. Xiayire Xiaokaiti, Dr. Yuki Matsumoto, Dr. Shuichi Matsumura, and Dr. Hitomi Hongo for their help with the Japanese wolf studies.

Competing Interests The author has no relevant financial or non-financial interests to disclose.

References

Bergström A, Frantz L, Schmidt R, Ersmark E, Lebrasseur O, Girdland-Flink L, Lin AT, Storå J, Sjögren K-G, Anthony D (2020) Origins and genetic legacy of prehistoric dogs. Science 370:557–564

Bergström A, Stanton DWG, Taron UH, Frantz L, Sinding MS, Ersmark E, Pfrengle S, Cassatt-Johnstone M, Lebrasseur O, Girdland-Flink L et al (2022) Grey wolf genomic history reveals a dual ancestry of dogs. Nature 607:313–320

Botigué LR, Song S, Scheu A, Gopalan S, Pendleton AL, Oetjens M, Taravella AM, Seregély T, Zeeb-Lanz A, Arbogast R-M (2017) Ancient European dog genomes reveal continuity since the early Neolithic. Nat Commun 8:1–11

Fan Z, Silva P, Gronau I, Wang S, Armero AS, Schweizer RM, Ramirez O, Pollinger J, Galaverni M, Del-Vecchyo DO (2016) Worldwide patterns of genomic variation and admixture in gray wolves. Genome Res 26:163–173

Frantz LA, Mullin VE, Pionnier-Capitan M, Lebrasseur O, Ollivier M, Perri A, Linderholm A, Mattiangeli V, Teasdale MD, Dimopoulos EA (2016) Genomic and archaeological evidence suggest a dual origin of domestic dogs. Science 352:1228–1231

Freedman AH, Gronau I, Schweizer RM, Ortega-Del Vecchyo D, Han E, Silva PM, Galaverni M, Fan Z, Marx P, Lorente-Galdos B (2014) Genome sequencing highlights the dynamic early history of dogs. PLoS Genet 10:e1004016. https://doi.org/10.1371/journal.pgen.1004016

Gojobori J, Arakawa N, Xiaokaiti X, Matsumoto Y, Matsumura S, Hongo H, Ishiguro N, Terai Y* (2024) Japanese wolves are most closely related to dogs and share DNA with East Eurasian dogs. Nat Commun 15(1):1680. https://doi.org/10.1038/s41467-024-46124-y

Inukai T (1982) Eradication of Ezo wolf by humans. Zenshu Nihon Doubutsushi 5:11–15

Ishiguro N, Inoshima Y, Shigehara N (2009) Mitochondrial DNA analysis of the Japanese wolf (*Canis lupus hodophilax* Temminck, 1839) and comparison with representative wolf and domestic dog haplotypes. Zool Sci 26:765–770

Ishiguro N, Inoshima Y, Yanai T, Sasaki M, Matsui A, Kikuchi H, Maruyama M, Hongo H, Vostretsov YE, Gasilin V (2016) Japanese wolves are genetically divided into two groups based on an 8-nucleotide insertion/deletion within the mtDNA control region. Zool Sci 33:44–49

Ishiguro N, Matsumura S, Terai Y, Hongo H (2021) Mystery of the Japanese wolf specimens left by Siebold, called wolf and yamainu. Nihon Jyuui Gakkai 74:389–395

Kishida, K. (1931). Notes on the Yesso wolf. Lansania 3:72–75.

Koepfli KP, Pollinger J, Godinho R, Robinson J, Lea A, Hendricks S, Schweizer RM, Thalmann O, Silva P, Fan Z, Yurchenko AA, Dobrynin P, Makunin A, Cahill JA, Shapiro B, Álvares F, Brito JC, Geffen E, Leonard JA, Helgen KM, Johnson WE, O'Brien SJ, Valkenburgh BV, Wayne RK (2015) Genome-wide evidence reveals that African and Eurasian golden jackals are distinct species. Curr Biol 25:2158–2165

Larson G, Karlsson EK, Perri A, Webster MT, Ho SY, Peters J, Stahl PW, Piper PJ, Lingaas F, Fredholm M (2012) Rethinking dog domestication by integrating genetics, archeology, and biogeography. Proc Natl Acad Sci 109:8878–8883

Leathlobhair MN, Perri AR, Irving-Pease EK, Witt KE, Linderholm A, Haile J, Lebrasseur O, Ameen C, Blick J, Boyko AR (2018) The evolutionary history of dogs in the Americas. Science 361:81–85

Loog L, Thalmann O, Sinding MHS, Schuenemann VJ, Perri A, Germonpré M, Bocherens H, Witt KE, Samaniego Castruita JA, Velasco MS (2020) Ancient DNA suggests modern wolves trace their origin to a late Pleistocene expansion from Beringia. Mol Ecol 29:1596–1610

Matsumura S, Inoshima Y, Ishiguro N (2014) Reconstructing the colonization history of lost wolf lineages by the analysis of the mitochondrial genome. Mol Phylogenet Evol 80:105–112

Matsumura S, Terai Y, Hongo H, Ishiguro N (2021) Analysis of the mitochondrial genomes of Japanese wolf specimens in the Siebold collection, Leiden. Zool Sci 38:60–66

Niemann J, Gopalakrishnan S, Yamaguchi N, Ramos-Madrigal J, Wales N, Gilbert MTP, Sinding MHS (2021) Extended survival of Pleistocene Siberian wolves into the early 20th century on the Island of Honshū. iScience 24:101904. https://doi.org/10.1016/j.isci.2020.101904

Perri AR, Mitchell KJ, Mouton A, Álvarez-Carretero S, Hulme-Beaman A, Haile J, Jamieson A, Meachen J, Lin AT, Schubert BW (2021) Dire wolves were the last of an ancient New World canid lineage. Nature 591:87–91

Savolainen P, Zhang Y, Luo J, Lundeberg J, Leitner T (2002) Genetic evidence for an east Asian origin of domestic dogs. Science 298:1610–1613

Segawa T, Yonezawa T, Mori H, Kohno A, Kudo Y, Akiyoshi A, Wu J, Tokanai F, Sakamoto M, Kohno N, Nishihara H (2022) Paleogenomics reveals independent and hybrid origins of two morphologically distinct wolf lineages endemic to Japan. Curr Biol 32:2494–2504

Shannon LM, Boyko RH, Castelhano M, Corey E, Hayward JJ, McLean C, White ME, Abi Said M, Anita BA, Bondjengo NI (2015) Genetic structure in village dogs reveals a central Asian domestication origin. Proc Natl Acad Sci 112:13639–13644

Shigehara N, Hongo H (2000) Ancient remains of Jomon dogs from neolithic sites in Japan. BAR Int Ser 889:61–70

Sinding M-HS, Gopalakrishnan S, Ramos-Madrigal J, de Manuel M, Pitulko VV, Kuderna L, Feuerborn TR, Frantz LA, Vieira FG, Niemann J (2020) Arctic-adapted dogs emerged at the Pleistocene–Holocene transition. Science 368:1495–1499

Skoglund P, Ersmark E, Palkopoulou E, Dalén L (2015) Ancient wolf genome reveals an early divergence of domestic dog ancestors and admixture into high-latitude breeds. Curr Biol 25:1515–1519

Temminck CJ (1839) Over de kennis en de verbreiding der Zoogdieren van Japan. Tijdschr Natuurl Gesch Physiol 5:273–293

Thalmann O, Shapiro B, Cui P, Schuenemann VJ, Sawyer SK, Greenfield D, Germonpré MB, Sablin MV, López-Giráldez F, Domingo-Roura X (2013) Complete mitochondrial genomes of ancient canids suggest a European origin of domestic dogs. Science 342:871–874

van der Valk T, Pečnerová P, Díez-del-Molino D, Bergström A, Oppenheimer J, Hartmann S, Xenikoudakis G, Thomas JA, Dehasque M, Sağlıcan E (2021) Million-year-old DNA sheds light on the genomic history of mammoths. Nature 591:265–269

von Holdt BM, Pollinger JP, Lohmueller KE, Han E, Parker HG, Quignon P, Degenhardt JD, Boyko AR, Earl DA, Auton A (2010) Genome-wide SNP and haplotype analyses reveal a rich history underlying dog domestication. Nature 464:898–902

Wang G-D, Zhai W, Yang H-C, Wang L, Zhong L, Liu Y-H, Fan R-X, Yin T-T, Zhu C-L, Poyarkov AD (2016) Out of southern East Asia: the natural history of domestic dogs across the world. Cell Res 26:21–33

Willerslev E, Cooper A (2005) Ancient DNA. Proc R Soc B Biol Sci 272:3–16

Xiaokaiti X, Sato T, Kasai K, Machida K, Yamazaki K, Yamaji N, Kikuchi H, Gojobori J, Hongo H, Terai Y, Gakuhari T (2023) The history of ancient Japanese dogs revealed by mitogenomes. Anthropol Sci 132(1):1–11. https://doi.org/10.1537/ase.230617

Chapter 5
Origin and Relationships Among Native Japanese Horse Breeds

Teruaki Tozaki

Abstract Currently, eight native horse breeds—the Hokkaido, Kiso, Misaki, Noma, Taishu, Tokara, Miyako, and Yonaguni—are bred in Japan. Native Japanese breeds are believed to have originated from Mongolian horses imported from the Korean Peninsula during fourth–fifth century. The phylogenetic relationships among these breeds are not well elucidated. This chapter first introduces the current theory behind the origin of native Japanese horses and then discusses the genetic diversity of native Japanese horses using genome-wide genetic markers, including autosomal chromosomes, the mitochondria genome (maternally inherited), and the Y chromosome (paternally inherited), and their origin. Finally, a synopsis of the results is used to propose an updated theory on the origin of native Japanese horses.

Keywords Breed history · Equine · Equus caballus · Hokkaido horse · Single nucleotide polymorphism

Abbreviations

HAN	Hanoverian Horse
PCA	Principal component analysis
PT	Paint Horse
QT	Quarter Horse
SNP	Single nucleotide polymorphisms
SZWB	Swiss Warm Blood

T. Tozaki (✉)
Genetic Analysis Department, Laboratory of Racing Chemistry, Utsunomiya, Tochigi, Japan
e-mail: ttozaki@lrc.or.jp

N. Osada et al. (eds.), *Phylogeographic History of Plants and Animals Coexisting with Humans in Asia*, Evolutionary Studies,
https://doi.org/10.1007/978-981-97-6887-5_5

5.1 Introduction

The theories underlying the origin of indigenous Japanese horse breeds primarily include the "two-wave immigration hypothesis" (Hayashida 1958) and the "single immigration hypothesis" (Nozawa et al. 1998).

Hayashida (1958) classified native Japanese horses into small- and medium-sized horses based on the difference in withers height (estimated using excavated bones of individual horses found at archeological sites) to explain the two-wave immigration hypothesis, whereby small-sized horses immigrated from the south and established themselves on the Japanese Islands around the late Jomon period (ca 18,000–300 BC), while medium-sized horses were imported to Japan via the Korean Peninsula and bred in Japan during the Kofun period (ca 300–600 AD). However, bones excavated from shell mounds of the Jomon period have been found to consist of individuals that died at a later date, thus disturbing the evidence.

Nozawa et al. (1998) claimed that the classification of lineages based on withers height is inconsistent with the phylogenetic relationships based on the statistical and genetic characteristics of blood-group protein polymorphisms in different breeds. Consequently, they proposed the single immigration hypothesis, whereby one original horse population was introduced. Blood-group protein polymorphisms were analyzed using objective indicators that do not depend on the subjectivity of the evaluator; however, compared to DNA genotypes, there are fewer polymorphic loci. Therefore, it is difficult to identify the origins and confirm the phylogenetic relationships between Japanese breeds. Nevertheless, with a small number of DNA microsatellites, Tozaki et al. (2003), also rejected the two-wave immigration hypothesis but found different genetic relationships between breeds compared with those proposed by Nozawa et al. (1998).

Based on a recent study by Tozaki et al. (2019), this chapter investigates the origin and phylogenetic relationships of eight native Japanese horse breeds relative to 32 international horse breeds using a broad range of single nucleotide polymorphisms (SNPs). Although native Japanese breeds are believed to have originated from ancient Mongolian horses imported from the Korean Peninsula, the phylogenetic relationships among these breeds are not well elucidated. In this chapter, a synopsis of the results is used to propose an updated theory on the origin of native Japanese horses.

5.2 Characteristics of Present-Day Native Japanese Horses

Eight horse breeds are currently recognized to be native to Japan by the Japan Equine Affairs Association (Fig. 5.1). These include the Hokkaido (*Hokkaido washuba* or *dosanko*), Kiso (*Kiso uma*), Noma (*Noma uma*), Taishu (*Taishu ba*), Misaki (*Misaki uma*), Tokara (*Tokara uma*), Miyako (*Miyako uma*), and Yonaguni (*Yonaguni uma*) breeds. The withers height of all native Japanese horses is <147 cm,

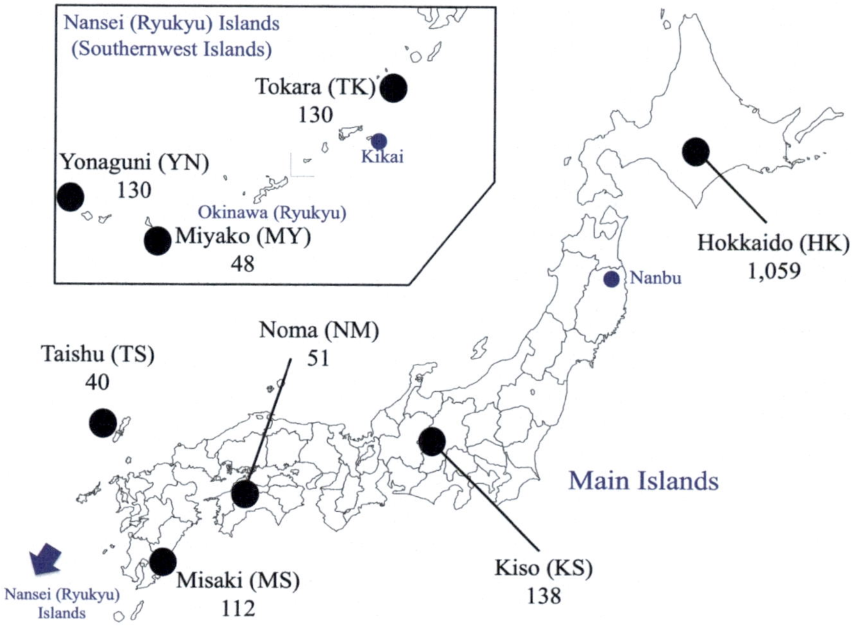

Fig. 5.1 Locations of the native Japanese Hokkaido, Kiso, Misaki, Noma, Taishu, Tokara, Miyako, and Yonaguni breeds. Numbers indicate population size as determined in 2017 (Tozaki et al. 2019)

and thus, they are classified as ponies. The characteristics of native Japanese horses are described below.

1. Hokkaido (*Hokkaido washuba* or *dosanko*)

   The Hokkaido is bred in the Hokkaido Prefecture (Amano et al. 2020); considering its withers height of 125–135 cm and body weight of 350–400 kg, it is a relatively large breed among native Japanese horses. The coat color varies and includes bay, buckskin, palomino, cremello, and roan. Although classified as a pony, Hokkaido horses can carry 110–220 kg. The breed has a sturdy constitution, which evolved as part of a primitive lifestyle under harsh natural conditions, with high physical strength for running. Many individuals possess the genetic ability to perform "pace," a two-beat gait where the horse moves both legs on the same side of the body (Amano et al. 2018; Saito et al. 2023).
2. Kiso (*Kiso uma*)

   The Kiso is primarily bred in the Nagano Prefecture (Takasu et al. 2011, 2012). It is a relatively large breed among native Japanese horses, with a mean withers height of 136 cm for males and 133 cm for mares, and a body weight of 350–420 kg. It is currently registered as a Kiso breed (purebred) and Kiso line (improved breed), with a dark dorsal stripe often observed in pure breeds. The Kiso has short legs and a long wide torso; its hind legs appear X-shaped, and its outward-pointed hooves are effective for firm footing on slopes, making movement on mountains easier. The coat is mostly bay-colored (Nakamura et al. 2019), which is due to most breeding stallions in the small remaining population being bay-colored.

3. Noma (*Noma uma*)

   The Noma is bred in Imabari City, Ehime Prefecture. It is a small breed with a withers height of 110–120 cm. Its coat color is primarily chestnut and bay. During the Edo period (1600–1868), the feudal lord of the Iyo-Matsuyama domain Sadayuki Matsudaira commissioned farming families from the Noma countryside within the Imabari clan territory to breed horses. In addition to covering their breeding expenses, families were given a cash bonus for horses larger than the specified size of approximately 121 cm, which stimulated Noma production. Horses smaller than the specified size were given to farming families free of charge instead of paying a breeding fee. Mating of these smaller horses gave rise to the present-day small-sized Noma.

4. Taishu (*Taishu ba*)

   The Taishu is bred in Tsushima, Nagasaki Prefecture (Tezuka et al. 2019; Yoshihara et al. 2022). Individuals range from small- to medium-sized, with withers height of 107–136 cm (mean height: 127 cm for males and 125 cm for mares). While the original coat color is considered to be black, the coat color of most present-day horses is bay, followed by black and chestnut. These horses have narrow chests and thin necks because they were bred for work on mountain paths, and these traits enabled easy transport of children and smaller adults.

5. Misaki (*Misaki uma*)

   The Misaki is bred in its wild state with almost no management. It has been allowed to freely graze in pastures throughout the year in Cape Toi, Miyazaki Prefecture over the last 300 years since the establishment of the breed (Kobayashi et al. 2019). It is approximately 130 cm at the withers and weighs approximately 300 kg. It is classified as a medium-sized horse among the native Japanese horses, with a solid build and a large head. However, compared with other native farm horses, it still shows characteristically thin legs observed in the riding horses of the Edo period. The coat color of the Misaki is often bay, and the ankles are black. Many individuals display the dark dorsal stripes observed in older horse breeds. Accordingly, these horses have qualities suited to their natural environment, such as a hardy constitution that can tolerate a plain diet. Further, their hind quarters have developed for navigating a hilly environment. *Misaki uma* lives in harems comprising one male, several females, and foals. Young males without harems form herds comprising only of males.

6. Tokara (*Tokara uma*)

   The Tokara used to be bred in Takarajima of the Tokara archipelago (Toshima village, Kagoshima District, Kagoshima Prefecture) and Kikaijima Island of the Amami Islands of Kagoshima Prefecture (Senokuchi et al. 2018). These horses are currently bred on the Iriki farm, Mount Kaimondake, Nakanoshima, and Kagoshima Prefecture. With withers height of 100–120 cm, the Tokara is the smallest class of horse among native Japanese horses. Its coat is mostly bay. The Tokara was primarily used for farming, transportation, and sugarcane processing.

7. Miyako (*Miyako uma*)

   The Miyako is bred on Miyako Island (*Miyakojima*) in the Okinawa Prefecture (Senju et al. 2017a). It is a small horse with a withers height of approximately 120 cm. The coat color is primarily bay. Being from a remote island, these horses have been maintained as pure breeds. They are warm and obedient in temperament, take to the owner well, and tolerate a plain diet and hard labor; therefore, they are used in farming and to carry heavy loads. Furthermore, their hooves are large and hard, making them suitable for coral stone roads and farming in sugarcane fields with thin soils. The breed played an active role in establishing sugarcane farming on Miyako Island in the Meiji period (1868–1912).

8. Yonaguni (*Yonaguni uma*)

   The Yonaguni is bred on Yonaguni Island (*Yonagunijima*) of the Okinawa Prefecture (Senju et al. 2017b). It is a small horse with withers height of 110–120 cm. The coat color is primarily bay. Having inhabited the westernmost remote island of Japan, these horses did not mate with other breeds, and selective breeding was not performed. In 1939, the stallion control law was implemented to promote breeding among native Japanese and Western breeds for horse improvement; however, the Yonaguni Island was excluded from the law-enforcement area, and was thus, devoid of crossbreeding.

5.3 Population Size and Genetic diversity of Native Japanese Horses

The Hokkaido and Kiso populations have maintained relatively high genetic diversity compared with that of other native Japanese and overseas breeds; however, the other six native species have lower genetic diversity and are in danger of losing breed continuity. This tendency is particularly evident in the Tokara, although this breed is managed under natural pasturage conditions, under which good levels of health can be maintained. Therefore, breeding and reproductive management are likely to be successful, regardless of the extreme loss of genetic diversity.

Native Japanese breeds have relatively poor overall genetic diversity compared with overseas breeds (Table 5.1), for which I propose the following two hypotheses: First, a limited number of individual horses were initially introduced throughout Japan; thus, the original level of genetic diversity was very low. Second, although various breeds were initially introduced into Japan, motorization along with industrialization caused an extreme reduction in the equine livestock headcount and the associated diversity. Considering that the Hokkaido remains relatively diverse, the second hypothesis may better explain the poor level of diversity among other native Japanese breeds.

For the Miyako and Kiso breeds, a decline in the population size corresponding with the use of fewer individuals caused a decrease in the genetic diversity (Masuda et al. 2023; Nakamura et al. 2020). These findings indicate the importance of maintaining genetic diversity by breeding horses outside of pure-bred lines.

Table 5.1 Diversity statistics for the eight Japanese breeds, the Japanese Thoroughbred, and previously studied breeds (Petersen et al. 2013), based on data from 20,287 loci (Tozaki et al. 2019)

		Individual Inbreeding (F)							
Breed name (abbreviation)		Min	Max	Mean	StDev	N	Proportion polymorphic	Expected heterozygosity	St Dev (H_e)
Hokkaido	HK	0.096	0.334	0.169	0.040	42	0.864	0.267	0.181
Kiso	KS	0.084	0.247	0.174	0.045	23	0.792	0.258	0.191
Misaki	MS	0.259	0.465	0.357	0.050	21	0.549	0.200	0.206
Miyako	MY	0.087	0.546	0.235	0.093	35	0.660	0.230	0.202
Noma	NM	0.146	0.354	0.254	0.046	24	0.604	0.225	0.206
Tokara	TK	0.379	0.598	0.456	0.057	24	0.437	0.158	0.200
Taishu	TS	0.075	0.442	0.228	0.114	25	0.743	0.255	0.193
Yonaguni	YN	0.123	0.467	0.345	0.082	24	0.667	0.206	0.199
Thoroughbred – Japan	TB_JP	−0.023	0.123	0.067	0.029	54	0.878	0.292	0.179
Thoroughbred – UK	TB_Eur	0.007	0.109	0.062	0.027	18	0.844	0.295	0.185
Thoroughbred – US	TB_US	−0.009	0.105	0.058	0.028	18	0.851	0.298	0.182
Thoroughbred – all	TB	−0.023	0.123	0.064	0.028	90	0.894	0.296	0.176
Akhal Teke	AH	0.010	0.224	0.075	0.053	20	0.862	0.301	0.175
Andalusian	AND	0.042	0.316	0.120	0.069	18	0.890	0.302	0.175
Arabian	ARR	0.038	0.316	0.113	0.060	24	0.887	0.297	0.178
Belgian	BEL	0.086	0.189	0.144	0.027	30	0.878	0.274	0.181
Clydesdale	CLYD	0.139	0.346	0.288	0.051	24	0.734	0.229	0.195
Caspian	CSP	−0.005	0.137	0.053	0.044	15	0.772	0.299	0.179
Exmoor	EX	0.081	0.546	0.262	0.089	24	0.780	0.245	0.191
Fell	FELL	0.101	0.210	0.143	0.031	21	0.848	0.276	0.184
Finnhorse	FINN	0.050	0.140	0.078	0.018	27	0.932	0.295	0.169
Franches Montagnes	FM	0.031	0.224	0.102	0.047	19	0.880	0.291	0.178
French trotter	FT	0.028	0.136	0.077	0.028	17	0.874	0.293	0.180

Hanoverian	HAN	−0.034	0.030	0.003	0.017	15	0.912	0.319	0.171
Icelandic	ICE	0.086	0.253	0.123	0.031	25	0.870	0.283	0.179
Miniature	MINI	0.070	0.339	0.114	0.053	25	0.914	0.288	0.173
Mangalaraga Paulista	MNGP	0.158	0.319	0.230	0.046	14	0.755	0.245	0.192
Mongolian	MON	−0.079	0.185	0.056	0.040	25	0.961	0.306	0.161
Morgan	MOR	0.010	0.306	0.092	0.081	43	0.949	0.305	0.165
New Forest pony	NFP	0.007	0.091	0.042	0.025	15	0.896	0.309	0.168
Norwegian Fjord	NORF	0.107	0.231	0.165	0.034	21	0.844	0.273	0.185
North Swedish horse	NSWE	0.109	0.235	0.164	0.036	19	0.823	0.270	0.185
Peruvian Paso	PERU	0.031	0.139	0.075	0.035	21	0.910	0.300	0.173
Percheron	PR	0.070	0.173	0.125	0.029	22	0.834	0.283	0.179
Puerto Rican Paso Fino	PRPF	0.008	0.311	0.127	0.065	20	0.875	0.281	0.180
Paint	PT	−0.028	0.057	0.001	0.020	25	0.958	0.323	0.160
Quarter horse	QH	−0.031	0.119	0.012	0.031	40	0.969	0.321	0.158
Saddlebred	SB	0.021	0.130	0.086	0.025	25	0.898	0.293	0.177
Shire	SH	−0.026	0.275	0.183	0.073	22	0.874	0.273	0.180
Shetland	SHET	0.136	0.406	0.218	0.061	27	0.831	0.260	0.186
Standardbred	STBD	0.056	0.166	0.111	0.036	25	0.876	0.286	0.182
Swiss WB	SZWB	−0.029	0.065	0.009	0.024	14	0.909	0.321	0.170
Tuva	TUVA	−0.013	0.145	0.050	0.044	15	0.913	0.313	0.165
Mean							0.833	0.277	0.181
Total						1005		0.322	0.144

5.4 Phylogenetic Relationships Between Native Japanese Horse Breeds

Tozaki et al. (2019) used approximately 9700 SNPs to examine the genetic relationships between individual horses using principal component analysis (PCA), a method in which dimensional reduction is performed on multidimensional data to visually characterize individuals according to breed based on the distribution of the data. The relationships among individuals are plotted with the first and second PCs as the axes in Fig. 5.2 (top) and the first and third PCs as the axes in Fig. 5.2 (bottom).

In PC1, the Hokkaido, Kiso, Noma, Misaki, and Taishu were grouped on the left, whereas the Tokara, Miyako, and Yonaguni were grouped on the right. This finding may explain why breeds of mainland Japan and those of the southwestern islands tend to differ genetically.

PC2 classified Tokara separately from the other breeds, which may be due to the different originating lines or extreme inbreeding. Considering the extremely poor genetic diversity of the Tokara compared with that of other breeds, extreme inbreeding may be more likely.

PC3 explains the genetic relationships between native horse breeds in mainland Japan (Hokkaido, Kiso, Noma, Misaki, and Taishu). The north and south of Japan are presented inversely in the top and bottom panels of Fig. 5.2 (bottom), in which each breed tended to show a consistent distribution.

The phylogenetic relationships of the native Japanese breeds are presented in Fig. 5.2 (bottom), based on the first and third PCs. The genetic relationships among the eight native breeds possibly reflect the local geographical distribution of breeding.

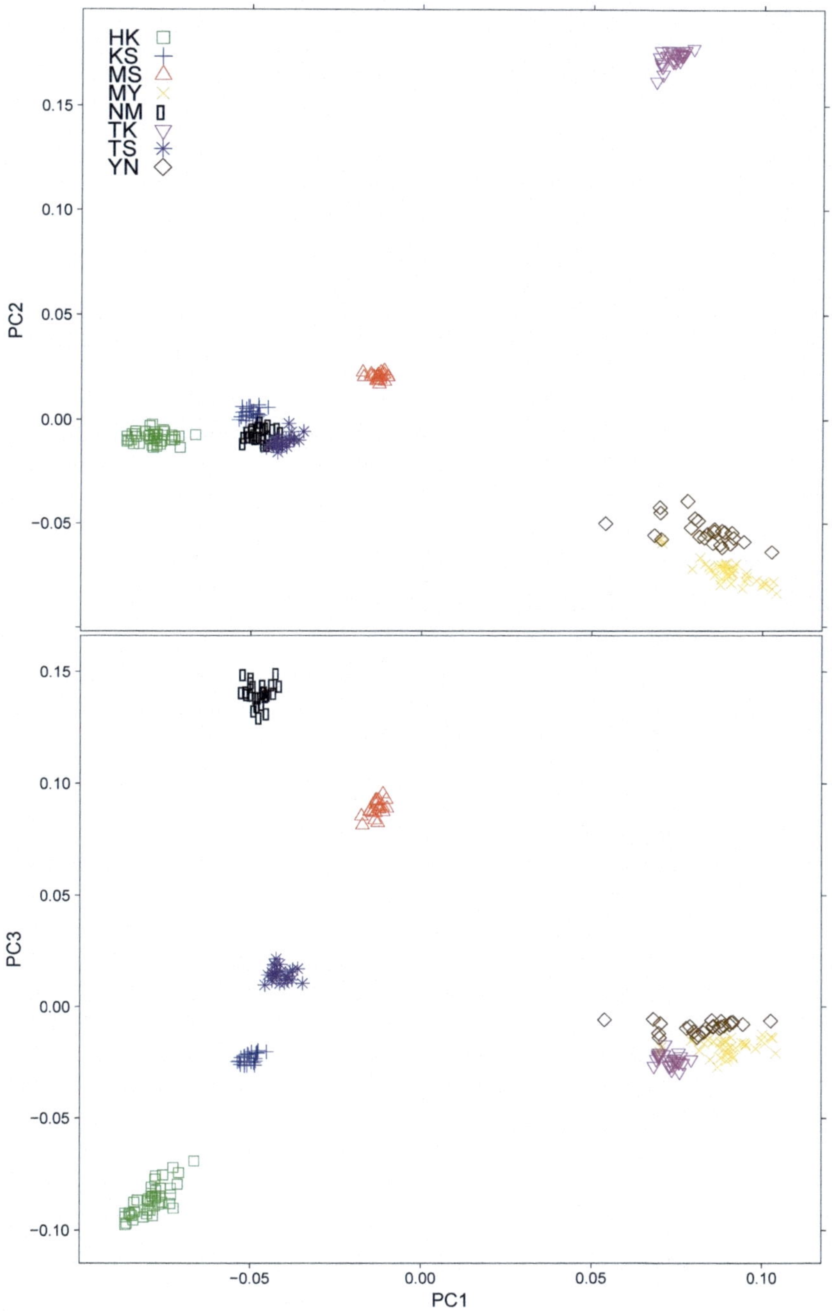

Fig. 5.2 Principal component biplots of the eight native Japanese breeds determined from 9700 SNP loci. (top) PC1 vs. PC2, (bottom) PC1 vs. PC3 (Tozaki et al. 2019)

5.5 Admixture Analyses

Admixture Among Native Japanese Breeds

Tozaki et al. (2019) conducted an admixture analysis to investigate the genetic structure of native Japanese horses. By setting a hypothetical number of ancestors (K) and examining the extent to which the populations share ancestral information at the individual or breed level, similarities in genetic structure can be elucidated, and relatedness between individuals or breeds can be inferred.

Figure 5.3 presents the results for the eight native Japanese breeds (218 horses). When K = 2 and 3, the Tokara, Miyako, and Yonaguni breeds in the southwestern islands are grouped into the same cluster; however, when K = 4, the Tokara is separated from the other breeds. When K = 7, the Miyako and Yonaguni breeds are grouped into the same cluster, and are inferred to have similar genetic structures. The Hokkaido, Kiso, and Taishu breeds are grouped in the same cluster up to K = 4; therefore, these three breeds are also inferred to possess the same genetic structure.

Admixture Among International Breeds

Tozaki et al. (2019) conducted an admixture analysis of 32 international breeds to investigate the degree of genetic variation among native Japanese horses (Fig. 5.4). When K = 4, the Tokara, Miyako, and Yonaguni of the southwestern Japanese islands showed distinctive clusters and were found to have a specific genetic structure compared to other breeds. When K = 29, the Miyako population was broadly divided into two ancestral groups or major reproductive populations.

Although the genetic structure was divided among native Japanese breeds (ancestral component), in international breeds, the same population included several ancestral components, and the difference between breeds was mainly apparent in the frequency of ancestral components. This is likely because individual breeds do not form extreme subpopulations. For example, when K = 30, the genetic structure and frequency were roughly the same for the Paint Horse (PT), Quarter Horse (QT), Swiss Warm Blood (SZWB), and Hanoverian Horse (HAN) in terms of structural composition. In contrast, native Japanese breeds appear to have a stronger tendency to form subpopulations.

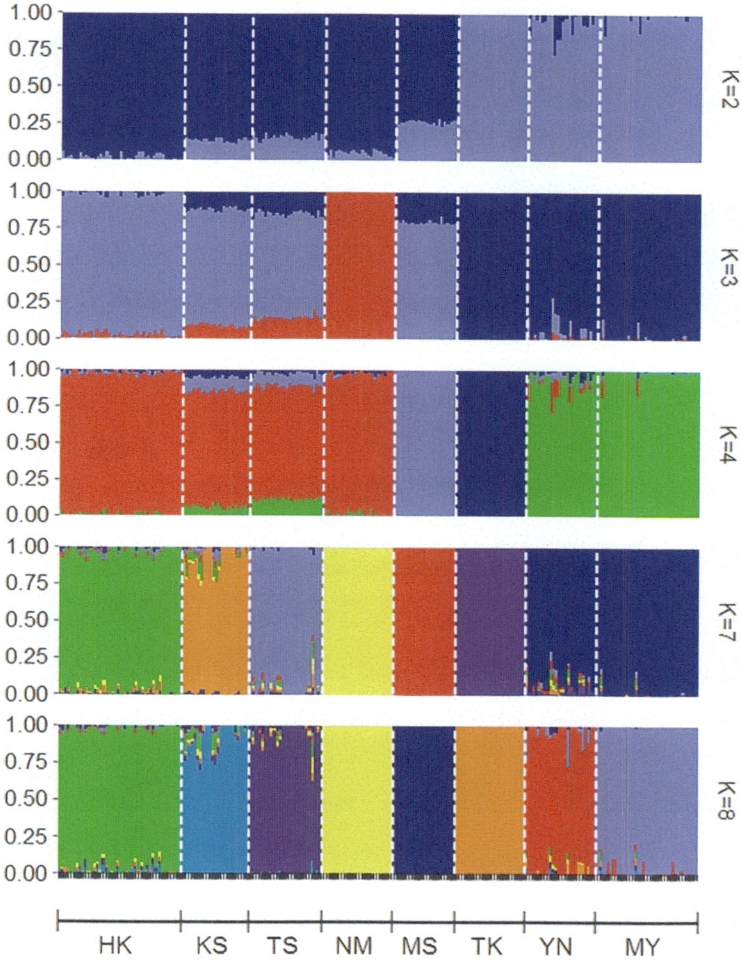

Fig. 5.3 Clustering output for five hypothetical K values for 218 native Japanese horses. Each individual is represented by one vertical line with the proportion of assignment to each cluster shown on the y-axis (Tozaki et al. 2019)

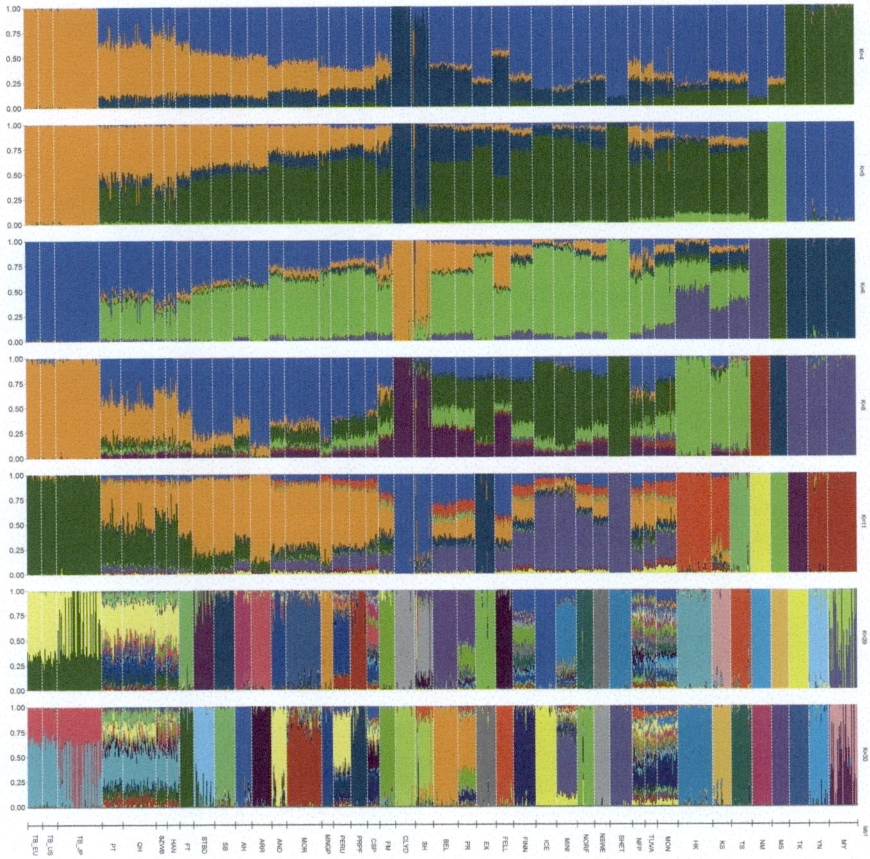

Fig. 5.4 Clustering output for 10 hypothetical K values for 1005 horses of 42 populations. Each individual is represented by one vertical line with the proportion of assignment to each cluster shown on the y-axis (Tozaki et al. 2019)

5.6 Phylogenetic Analyses

Phylogenetic Relationship Among International Breeds

PCA and ancestral group analysis were performed on the genetic data of individual horses to evaluate the similarities in genetic structure and distribution within and between breeds (Tozaki et al. 2019). The 40 breeds in the analysis were broadly classified into distinct lineages by phylogenetic analysis, such as the native Japanese, European, Southern European, and South American breeds and improved Thoroughbreds (Fig. 5.5).

The phylogenetic tree indicated that all native Japanese breeds have a single lineage linked to Mongolian horses. The European Fell ponies and larger Shire and

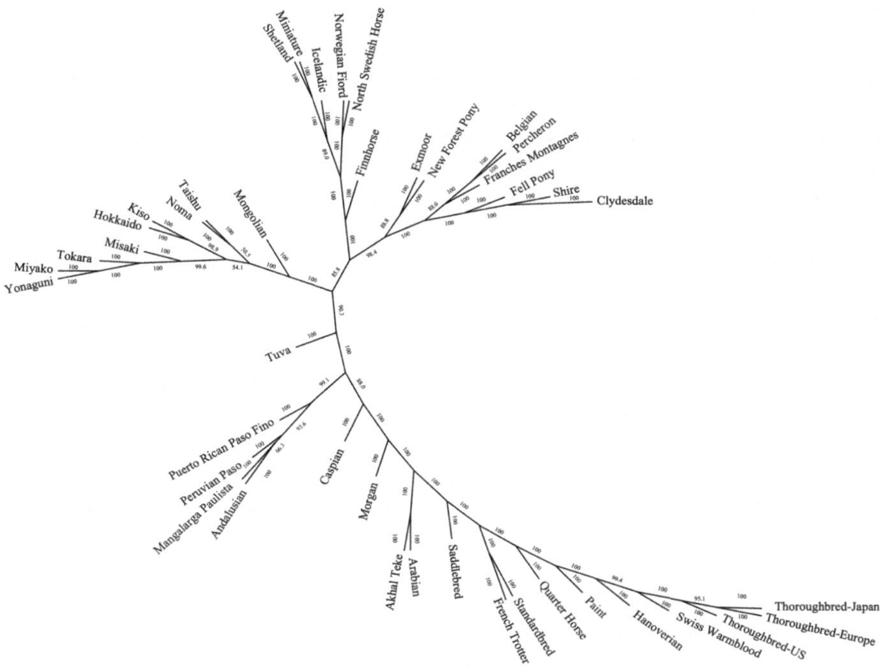

Fig. 5.5 Distance-based neighbor-joining tree calculated from SNP frequencies for 8 native Japanese breeds, Japanese Thoroughbreds, and 33 populations of the Equine Genetic Diversity Consortium. Bootstrap values (proportion of 1000 replicates) supporting each branch are noted (Tozaki et al. 2019)

Clydesdale horses (used to pull horse carts) were classified into the same lineage, regardless of breeding improvements made to alter the sizes of these horses. Furthermore, the Southern European (Andalusian) and South American breeds were likely classified into the same lineage because Southern European breeds were introduced into South America during the Age of Exploration.

Phylogenetic Relationship Between East Asian and Japanese Breeds

As shown in Fig. 5.5, native Japanese breeds originated from Mongolian horses, irrespective of their mainland or island distributions. Although the Tuva (Russian Federation; a breed from East Asia) was bred near the distribution of Mongolian horses, it has a slightly different genetic lineage, likely due to unique management and selective breeding practices.

The native Japanese breeds were classified into the Taishu and Noma groups, Hokkaido and Kiso groups, Misaki, and breeds of the southwestern Japanese islands

(Tokara, Miyako, and Yonaguni). The Taishu are bred in Tsushima, which is located halfway between the Korean Peninsula and Kyushu, which may have been used as a layover point for importing horses. Therefore, some genetic traces from the initially introduced horses may remain.

Little information exists in the literature or among local communities regarding the founder population of the Noma; however, the present results suggest that the breed may have originated from founder populations bred in the Tsushima or Kyushu regions.

Although the origin of the Hokkaido is unclear, various hypotheses exist. One suggests that fishermen of the Tohoku northern coast, who worked in Hokkaido in the summer, brought horses from Honshu, which subsequently became feral. One can further infer that the old Hokkaido originated from Nanbu horses (which are already extinct) in northern Tohoku. Further, the grouping of the Hokkaido and Kiso breeds may be explained by the introduction of Nanbu horses, of the Kai founder population, when horse breeding in the Kiso region first declined.

The Misaki breed formed when breeding and management were initiated by the Takanabe clan in 1697, after which the introduction of non-native horses was avoided. Therefore, they tend to be separate from the other breeds in the phylogenetic tree.

Based on the phylogenetic tree, breeds from the southwestern Japanese islands did not branch from nodes near the ancestral Mongolian horses. This contradicts the two-wave immigration hypothesis and supports the introduction of horses to Japan from a single origin.

Diversity and Phylogeny of the Maternal Line

Maternal inheritance was determined by examination of mitochondria (the cellular energy-producing organelles). The D-loop sequence in the mitochondrial genome does not encode functional genes and is full of polymorphisms; thus, it is suitable for the phylogenetic analysis of various life forms. The mitochondrial Eve theory has been investigated in many studies and suggests that based on differences in mitochondrial genome sequences, humankind originated in Africa. In horses, the mitochondrial genome exhibits extremely diverse haplotypes (mitochondrial genotypes), which is attributed to the domestication of mares in various regions.

Kakoi et al. (2007) investigated the phylogenetic relationships between native Japanese and international breeds by comparing the mitochondrial D-loop region. They observed that common haplotypes tend to be shared among all breeds; however, they also confirmed that some haplotypes were largely genetically isolated to a single breed. Therefore, while the maternal lines of native Japanese horse breeds vary, it is difficult to classify breed lines using maternal genetics alone.

Diversity and Phylogeny of the Paternal Line

Y chromosome markers are used to examine the paternal line. Although the horse Y chromosome has not been completely decoded, Kakoi et al. (2018) used partially decoded Y-chromosome markers to examine the diversity of the paternal line in 159 horses (stallions) of eight native Japanese breeds and identified three haplotypes (JHT-1, JHT-2, and JHT-3). Among these, JHT-3 was found in one Yonaguni individual, which was attributed to a spontaneous mutation; therefore, JHT-1 and JHT-2 were the main haplotypes in native Japanese breeds.

All horses of the Kiso and Misaki breeds carried JHT-2, while the remaining six breeds carried JHT-1. Therefore, from the perspective of paternal genetic structure, native Japanese horse breeds are classified into the JHT-1 (Hokkaido, Taishu, Noma, Tokara, Yonaguni, and Miyako) and JHT-2 groups (Kiso and Misaki).

Kakoi et al. (2018) identified JHT-1 and JHT-2 in native Chinese horses and claimed that most native Japanese breeds should carry these haplotypes. They suggested that only JHT-2 was observed in the Kiso and Misaki breeds because of the unique reproductive management of these breeds.

The Kiso population greatly declined during the Showa period (1926–1989) due to the Stallion Control Law, motorization, and because only one stallion was available for breeding for some time, which explains why only JHT-2 was observed in this breed (Kakoi et al. 2018). The small population of the Misaki was repeatedly mated since 1697, which generated only the JHT-2 haplotype. This could also explain why only the JHT-1 haplotype was observed in the Taishu, Noma, Tokara, Yonaguni, and Miyako breeds. However, a larger sample size is required to confirm the haplotypes carried in the Hokkaido breed.

In native Japanese breeds, there is little Y chromosome diversity, and it remains difficult to determine the origin and phylogeny of native Japanese breeds from the perspective of paternal inheritance.

5.7 Origins of Japanese Breeds

Origins of Mainland Japanese Breeds

As shown in the third PC (Fig. 5.2), the genetic relationships of the Hokkaido, Kiso, Taishu, Noma, and Misaki breeds correlated with the geographical relationships of their breeding regions (see Fig. 5.1). Admixture analysis revealed that the Hokkaido, Kiso, and Taishu breeds have a genetic structure similar to that of Mongolian breeds (Fig. 5.4). Thus, one can infer that Mongolian horses were introduced to Japan, and as horse production and breeding techniques spread to different regions with human migration, the genetic and geographical distributions gradually became more distinct.

According to the literature and local information, the breeding of Misaki horses began in 1697, after which crossbreeding was avoided, and they were kept in a half-feral state. Moreover, Noma horse breeding began in Imabari in 1635. Therefore, these two breeds may be genetically distinct even in mainland Japan.

Origins of Breeds from the Southwestern Japanese Islands

Although the origins of horses in the southwestern islands of Japan, including the main island of Okinawa, Miyako, and Yonaguni, are unclear from the PCA, ancestral group, and phylogenetic analyses; the native horses of these regions are presumed to be genetically related and stem from the same lineage (Senju et al. 2018).

The Miyako and Yonaguni islands were under the rule of the Ryukyu Kingdom when the Miyako and Yonaguni breeds were introduced, and were, therefore, bred by the same ruler. Therefore, these breeds are expected to be closely related. The Tokara breed is native to Takara Island, which is located at the southern tip of the Tokara archipelago, and was brought from Kikai Island in 1897. While the origin of the horses on Kikai Island is unclear, based on their historical distributions, we cannot rule out a relationship with native horses of the Okinawa region.

Native horses from the southwestern islands of Japan (Tokara, Miyako, and Yonaguni) are thought to be genetically isolated from mainland Japan because of inbreeding and genetic drift. The reductions in genetic diversity and common alleles associated with the decline in livestock headcount are attributed to a decrease in similarity with other breeds. As mentioned above, the genetic diversity of the Tokara, Miyako, and Yonaguni breeds tended to be lower than that of mainland Japanese breeds.

Native Japanese Breeds Originating from Domestic Horses

The oldest traces of horse domestication are found in the ruins of the ancient Botai culture, which existed approximately 5500 years ago (Gaunitz et al. 2018), including horse-riding harnesses and traces of horse milk left in earthenware. Genome data and phylogenetic analyses revealed that these ancient horses were not of the same breed as modern domestic horses but rather related to the endangered wild Przewalski's horses (Mongolian wild horses). Instead, the ancestors of the modern domesticated horses originated from a breed that inhabited the steppe of modern-day southern Russia (around the Volga and Don rivers) approximately 4200 years ago (Librado et al. 2021).

In the two-wave immigration hypothesis, native horses of the southwestern islands of Japan are considered to have originated from a group of wild horses that moved northward from the south in the Jomon period. However, the modern-day

Tokara, Miyako, and Yonaguni breeds are more closely related to the domestic horses introduced from southern Russia, which rules out the two-wave hypothesis.

Native Japanese Horses Originating from a Single Lineage

Considering the results of the lineage classification between individuals and breeds and the overall degree of genetic diversity, it is conceivable that native Japanese horses originated from the native Mongolian horse population and subsequently spread throughout Japan, forming geographically distinct subpopulations because of reproductive isolation, which led to the eight native breeds of present-day Japan.

Horses that are native to the southwestern islands of Japan (Tokara, Miyako, and Yonaguni) have a similar genetic structure and tend to be relatively isolated from mainland Japanese breeds; however, relative to international breeds, the island breeds do not differ substantially in their genetics from the native horse lineages of mainland Japan. The present-day Tokara, Miyako, and Yonaguni undoubtedly originated from a mainland Japanese lineage. Both native Mongolian and Japanese horse breeds belong to the present-day domestic horse (East Asian) lineage and are thought to differ genetically from the European and Middle Eastern lineages.

Kofun Period Immigration and the Introduction of Horses and Related Skills

A three-stage immigration model was recently proposed for establishing the Japanese human population (Jinam et al. 2021). In the first stage, hunter-gatherers migrated to Japan from the Paleolithic to the late Jomon period (approximately 40,000–3000 years ago). In the second stage, rice cultivation was introduced to the coast of Northern Kyushu from the continent via the southern part of the Korean Peninsula during the Yayoi period (approximately 1000 BC to 201–300 AD). In the newly proposed third stage, earthenware artisans and iron workers settled in Japan during the Kofun period (201–300 AD onward).

Horses, along with horse-riding equipment (e.g., harnesses) and breeding techniques, were introduced to Japan during the fourth and fifth century. Training is required to ride horses, and expert skills are required to manufacture harnesses. Therefore, Japanese horses as well as riding and breeding-related skills may be introduced by the third stage immigrant people of the three-stage immigration model.

Although there were no harnesses buried in the ancient tombs of the early Kofun period, almost all tombs of the fifth-century mid-Kofun period contained harnesses. Furthermore, the moats surrounding tombs were often burial sites for sacrificed horses; this was possibly customary for skilled settlers.

5.8 Conclusion

By analyzing eight native Japanese breeds and 32 international breeds using genome-wide SNPs, I found that native Japanese horses originated from the native Mongolian horse population and spread throughout Japan after being introduced, forming geographically distinct subpopulations as a result of natural reproduction and artificial breeding, which led to the establishment of the eight native breeds of present-day Japan. However, the research introduced here did not evaluate genetic crossbreeding with horses in the Asian continent such as China, Korea, etc. (Kim et al. 2018; Ma et al. 2019; Srikanth et al. 2019). Therefore, future studies on the genetic structure and origins of native Japanese horses should consider the influence of horses from East Asia.

Competing Interests The author has no relevant financial or non-financial interests to disclose.

References

Amano T, Onogi A, Yamada F, Kawai M, Shirai K, Ueda J (2018) Genome-wide association mapping and examination of possible maternal effect for the pace trait of horses. Anim Genet 49:461–463

Amano T, Tozaki T, Takasu M, Onogi A, Yamada F, Kawai M, Ueda J (2020) Changes of sires in a breeding farm enables maintenance of DNA-level genetic variation in a produced herd of Hokkaido native horses. Anim Sci J 91:e13318

Gaunitz C, Fages A, Hanghøj K, Albrechtsen A, Khan M et al (2018) Ancient genomes revisit the ancestry of domestic and Przewalski's horses. Science 360:111–114

Hayashida S (1958) Ancestry of the native horses of Japan. Jpn J Zootech Sci 28:329–334. (in Japanese, with English summary)

Jinam TA, Kawai Y, Saitou N (2021) Modern human DNA analyses with special reference to the inner dual-structure model of Yaponesian. Anthrop Sci 129:3–11

Kakoi H, Tozaki T, Gawahara H (2007) Molecular analysis using mitochondrial DNA and microsatellites to infer the formation process of Japanese native horse populations. Biochem Genet 45:375–395

Kakoi H, Kikuchi M, Tozaki T, Hirota KI, Nagata SI, Hobo S, Takasu M (2018) Distribution of Y chromosomal haplotypes in Japanese native horse populations. J Equine Sci 29:39–42

Kim NY, Seong HS, Kim DC, Park NG, Yang BC, Son JK, Shin SM, Woo JH, Shin MC, Yoo JH, Choi JW (2018) Genome-wide analyses of the Jeju, thoroughbred, and Jeju crossbred horse populations using the high density SNP array. Genes Genomics 40:1249–1258

Kobayashi I, Akita M, Takasu M, Tozaki T, Kakoi H, Nakamura K, Senju N, Matsuyama R, Horii Y (2019) Genetic characteristics of feral Misaki horses based on polymorphisms of microsatellites and mitochondrial DNA. J Vet Med Sci 81:707–711

Librado P, Khan N, Fages A, Kusliy MA, Suchan T et al (2021) The origins and spread of domestic horses from the Western Eurasian steppes. Nature 598:634–640

Ma H, Wang S, Zeng G, Guo J, Guo M, Dong X, Hua G, Liu Y, Wang M, Ling Y, Ding X, Zhao C, Wu C (2019) The origin of a coastal indigenous horse breed in China revealed by genome-wide SNP data. Genes (Basel) 10:241

Masuda M, Nikadori A, Nikadori E, Uchida N, Takizawa Y, Ishimaru M, Yamamoto T, Yuasa R, Kugai H, Nagahama Y, Takasu M, Tozaki T (2023) Evaluation of genetic diversity using 31 microsatellites in Miyako horses. J Equine Sci 34:1–6

Nakamura K, Tozaki T, Kakoi H, Owada S, Takasu M (2019) Variation in the MC1R, ASIP, and MATP genes responsible for coat color in Kiso horse as determined by SNaPshot™ genotyping. J Vet Med Sci 81:100–102
Nakamura M, Tozaki T, Kakoi H, Nakamura K, Rajabi-Toustani R, Ohba Y, Matsubara T, Takasu M (2020) Decreased genetic diversity in Kiso horses revealed through annual microsatellite genotyping. J Vet Med Sci 82:503–540
Nozawa K, Shotake T, Ito S, Kawamoto Y (1998) Phylogenetic relationships among Japanese native and alien horses estimated by protein polymorphisms. J Equine Sci 9:53–69
Petersen JL, Mickelson JR, Cothran EG, Andersson LS, Axelsson J, Bailey E, Bannasch D, Binns MM, Borges AS, Brama P, da Câmara Machado A, Distl O, Felicetti M, Fox-Clipsham L, Graves KT, Guérin G, Haase B, Hasegawa T, Hemmann K, Hill EW, Leeb T, Lindgren G, Lohi H, Lopes MS, McGivney BA, Mikko S, Orr N, Penedo MC, Piercy RJ, Raekallio M, Rieder S, Røed KH, Silvestrelli M, Swinburne J, Tozaki T, Vaudin M, M Wade C, McCue ME (2013) Genetic diversity in the modern horse illustrated from genome-wide SNP data. PLoS One 8:e54997. https://doi.org/10.1371/journal.pone.0054997
Saito I, Nakamura K, Tozaki T, Hano K, Takasu M (2023) Genetic characterization of Japanese native horse breeds by genotyping variants that are associated with phenotypic traits. J Equine Sci 34:115–120
Senju N, Tozaki T, Kakoi H, Almunia J, Maeda M, Matsuyama R, Takasu M (2017a) Genetic characterization of the Miyako horse based on polymorphisms of microsatellites and mitochondrial DNA. J Vet Med Sci 79:218–223
Senju N, Tozaki T, Kakoi H, Shinjo A, Matsuyama R, Almunia J, Takasu M (2017b) Genetic diversity of the Yonaguni horse based on polymorphisms in microsatellites and mitochondrial DNA. J Vet Med Sci 79:425–431
Senju N, Tozaki T, Kakoi H, Matsuyama R, Nakamura K, Takasu M (2018) Genetic relationship between Miyako and Yonaguni horses native to Okinawa based on polymorphisms of microsatellites. J Equine Sci 29:87–90
Senokuchi A, Ishikawa S, Tozaki T, Takasu M, Kakoi H, Misumi K, Hobo S (2018) Genetic analyses for conservation of the traditional Tokara horse using 31 microsatellite markers. J Equine Sci 29:97–104
Srikanth K, Kim NY, Park W, Kim JM, Kim KD, Lee KT, Son JH, Chai HH, Choi JW, Jang GW, Kim H, Ryu YC, Nam JW, Park JE, Kim JM, Lim D (2019) Comprehensive genome and transcriptome analyses reveal genetic relationship, selection signature, and transcriptome landscape of small-sized Korean native Jeju horse. Sci Rep 9:16672
Takasu M, Hiramatsu N, Tozaki T, Kakoi H, Hasegawa T, Maeda M, Huricha, Kusuda S, Doi O, Murase T, Mukoyama H (2011) Population statistics and biological traits of endangered kiso horse. J Equine Sci 22:67–72
Takasu M, Hiramatsu N, Tozaki T, Kakoi H, Nakagawa T, Hasegawa T, Huricha, Maeda M, Murase T, Mukoyama H (2012) Genetic characterization of the endangered Kiso horse using 31 microsatellite DNAs. J Vet Med Sci 74:161–166
Tezuka A, Takasu M, Tozaki T, Nagano AJ (2019) Genetic analysis of Taishu horses on and off Tsushima Island: implications for conservation. J Equine Sci 30:33–40
Tozaki T, Takezaki N, Hasegawa T, Ishida N, Kurosawa M, Tomita M, Saitou N, Mukoyama H (2003) Microsatellite variation in Japanese and Asian horses and their phylogenetic relationship using a European horse outgroup. J Hered 94:374–380
Tozaki T, Kikuchi M, Kakoi H, Hirota K, Nagata S, Yamashita D, Ohnuma T, Takasu M, Kobayashi I, Hobo S, Manglai D, Petersen JL (2019) Genetic diversity and relationships among native Japanese horse breeds, the Japanese thoroughbred and horses outside of Japan using genome-wide SNP data. Anim Genet 50:449–459
Yoshihara T, Kikuchi M, Akiba Y, Horiguchi M, Takasu M, Tozaki T (2022) Genetic diversity analysis and parentage verification of Taishu horses using 31 microsatellites. J Equine Sci 33:63–69

Chapter 6
Molecular Phylogeographic Analyses Provide a New Perspective on the Evolutionary Origin and History of Japanese Native Chickens

Takahiro Yonezawa

Abstract Japanese native chickens (JNC) are the most diverse breed group worldwide, but several aspects of their origin and history have not yet been clarified. To determine the evolutionary relationships of JNCs with the world's native chickens, an overview of chicken domestication and the propagation of chickens on a global scale is provided. Subsequently, I focused on the origin of JNC in the Asian native chickens, a phylogenetic network on the basis of mitochondrial DNA from domestic chicken in continental Southeast Asia and East Asia was inferred and it revealed a geographic gradient from Southeast Asia to North China. Populations of the historical periphery of the Sinocultural sphere, including Japan and Korea, are derived from North Chinese population. This suggests that JNCs originated in North China or Korean Peninsula. Some JNC breeds, especially the cockfighting breed *Shamo*, are thought to have originated in Thailand, but these breeds were more closely related to the Chinese population compared with the Southeast Asian population.

Keywords Mitochondrial genome · Time tree · Jidori · Shamo · Shokoku

Abbreviations

ISEA Islands of Southeast Asia
JNC Japanese native chicken
OTU Operational taxonomic unit

T. Yonezawa (✉)
Graduate School of Integrated Sciences for Life, Hiroshima University, Hiroshima, Japan
e-mail: tyonezaw@hiroshima-u.ac.jp

N. Osada et al. (eds.), *Phylogeographic History of Plants and Animals Coexisting with Humans in Asia*, Evolutionary Studies,
https://doi.org/10.1007/978-981-97-6887-5_6

6.1 Introduction

There are currently more than 31 billion chickens on our planet (FAOSTAT: https://www.fao.org/faostat/en/#home), and they are kept in every human settlement of the world (except for Antarctica and Vatican City), transcending ethnicities, religions, and national borders (Lawler 2014). Chickens are an essential part of today's mass consumption society. Chickens have played a key role in human history not only as a source of animal protein but also in sociocultural contexts (Herrera et al. 2020). Because of their small body size, chickens have spread with humans via migration and trade. For example, Austronesians left Taiwan in approximately 3000 BC, and their superior navigational skills allowed them to spread throughout the Pacific Ocean and even reach Madagascar. Austronesians reached Easter Island in easternmost Polynesia approximately 1000 years ago, and their migration was probably accompanied by chickens. A phylogenetic tree on the basis of the mitochondrial genomes of native Pacific chickens is highly consistent with the migratory patterns of Austronesians (Yonezawa et al. 2022). Clarifying the phylogeographic history of chickens is key for understanding patterns of material exchange between cultural spheres.

In this chapter, I focus on the history of Japanese native chickens and outline updated findings from molecular phylogeographic analyses. Japanese native chickens comprise a group of breeds that were introduced to Japan before the Meiji Restoration (1868) or were originally bred in Japan. There is no consensus on the exact number of breeds of Japanese native chickens, but Tsudzuki (2003) suggests that there are approximately 40 known breeds. Considering that the total number of chicken breeds in the world is between 300 and 400, Japanese native chickens comprise a large proportion of the world's chicken breeds. Zooarchaeological evidence (Eda et al. 2023) suggests that chickens have been used in Japan since the Yayoi period (tenth century BC–third century). However, since most of the known chicken bones from the Yayoi period are male, such a significant sex ratio bias in this period affected the maintenance of their populations in Japan; thus, chickens might have been repeatedly introduced to Japan from the Asian continent (Eda 2016).

This complexity has contributed to the longstanding lack of a uniform perspective on the origin and history of Japanese native chickens. Although several molecular genetic studies have been conducted (Oka et al. 2007; Osman et al. 2006; Hata et al. 2020), most of these studies have focused on the diversity and phylogenetic relationships within Japanese native chickens; however, the phylogenetic relationships of Japanese native chickens with native chickens worldwide, especially in Asia, have not been thoroughly studied. Here, I provide a summary of the origin and history of Japanese native chickens according to phylogeographical research of Asian native chickens.

6.2 The Wild Progenitor of Domestic Chickens

Before delving into the origin and history of Japanese native chickens, an overview of chicken domestication and the propagation of chickens on a global scale is provided. The wild progenitor of domestic chickens remained unclear for a long time. Chickens belong to the genus *Gallus* in the family Phasianidae include four extant wild species, red junglefowl (*G. gallus*), Ceylon junglefowl (*G. lafayetii*), gray junglefowl (*G. sonneratii*), and green junglefowl (*G. varius*). In "The Variation of Animals and Plants under Domestication, "Charles Darwin argued that red junglefowl might be the wild progenitor of domestic chickens (Darwin 1868). Various molecular genetic analyses have provided support for Darwin's hypothesis (Akishinonomiya et al. 1996; Miao et al. 2013; Lawal et al. 2020). Other *Gallus* species have also made genetic contributions to domestic chickens. For example, phenotypic polymorphism is observed in the shank color of domestic chickens. Although the legs of domestic chickens are often yellow, the legs of red junglefowl are black. Chicken shank color is known to be a Mendelian trait. Eriksson et al. (2008) showed that the chicken leg phenotype is controlled by the *BCDO2* locus and further demonstrated that the yellow shank color observed in domestic chickens is most likely derived from gray junglefowl. A whole-genome analysis conducted by Lawal et al. (2020) indicated that the main ancestor of the domestic chicken is red junglefowl, but three other *Gallus* species have also made significant genetic contributions to the domestic chicken. Chickens are thought to have been domesticated from red junglefowl and then improved by incorporating genes from other *Gallus* species during the breeding process.

6.3 Origin of Domestic Chickens and Their Global Propagation

So if the main ancestor of the chicken is the red junglefowls, in which region were they domesticated? Red junglefowl are widely distributed in the tropical rainforests of South Asia, continental Southeast Asia, and the islands of Southeast Asia (ISEA). They have been classified into five subspecies according to their geographic distributions and morphological characteristics; thus, clarification of the genetic relationships among domestic chicken and the red junglefowls subspecies can provide insights into regions in which chickens were domesticated. Akishinonomiya et al. (1996) demonstrated that the subspecies *spadiceus* from western continental Southeast Asia is closely related to the domestic chicken using mitochondrial D-loop sequences. This suggests that chicken domestication occurred in continental Southeast Asia, but the subspecies *murghi*, which is distributed in South Asia, and the subspecies *jabouillei*, which occurs in eastern continental Southeast Asia, were not included in this analysis. Subsequent studies with extensive taxon sampling (Liu et al. 2006; Kanginakudru et al. 2008) have shown that chicken domestication occurred not only in Southeast Asia (in the broad sense, including South China) but also in South Asia.

Currently, the mitochondrial DNA of domestic chickens/red junglefowl is classified into Haplogroups A–I (including domestic chickens) and V–Z (red junglefowl only) (Liu et al. 2006; Miao et al. 2013; Huang et al. 2018). Haplogroups A, B, and E are widely distributed, and individuals originating from various regions are intricately intermingled in the phylogenetic tree. Weak phylogeographic structure was observed in chicken mitochondrial DNA, and this increases the difficulty of clarifying the processes underlying the spread of domestic chickens on a global scale based on the relationships among individuals. This suggests that chickens were transported among regions at a high frequency. However, the composition of mitochondrial haplogroups in chickens varies widely among regions (Miao et al. 2013). Therefore, phylogenetic trees inferred at the population level might provide more insights into the evolutionary history of chicken domestication and propagation compared with trees inferred at the individual level.

I estimated the phylogenetic relationships among red junglefowl subspecies and domestic chickens of various regions using mitochondrial D-loop sequences from a global set of samples (Yonezawa and Sasaki 2016). The subspecies *bankiva* distributed in Java, Indonesia, diverges first, and followed by the Indonesian Sumatran population of the subspecies *gallus* (Fig. 6.1a). This finding is consistent with the results of previous studies (Akishinonomiya et al. 1996; Wang et al. 2020). Among red junglefowl on the Asian continent, the subspecies *murghi* of South Asia diverged first, and this was followed by the subspecies *jabouilei*, *gallus* (continental population), and *spadiceus* of Southeast/East Asia. Furthermore, a branching pattern was observed, with East/Southeast Asian domestic chickens derived from the subspecies *spadiceus* and South/West Asian, European, North/East African, and Pacific/ISEA chickens derived from the subspecies *murghi*. Consistent with the results of previous studies (Liu et al. 2006; Kanginakudru et al. 2008), this indicates that chickens were domesticated in Southeast Asia and South Asia, which provides insights into their subsequent spread across the globe. Chickens domesticated in continental Southeast Asia spread widely throughout this region and further northward into East Asia. Domesticated populations in South Asia propagated westward and spread through West Asia to Europe and North Africa. Some lineages domesticated in South Asia also spread to East Africa, and others spread widely in the Pacific via ISEA.

Our phylogeographic analysis provides insights into the processes underlying the propagation of chickens across the globe. However, the above conclusion was derived from mitochondrial D-loop sequences (i.e., maternal history). The short length of mitochondrial D-loop sequences limits the amount of phylogenetic information that they provide. Indeed, our analysis using mitochondrial genome data indicates that ISEA/Pacific chickens likely originated in continental Southeast Asia (Yonezawa et al. 2022). Wang et al. (2020) performed a phylogeographic analysis using complete genome sequences of 863 domestic chickens/red junglefowl. In contrast to the results of analyses that have used mitochondrial DNA, analysis using nuclear genomes indicated that domestic chickens were derived exclusively from

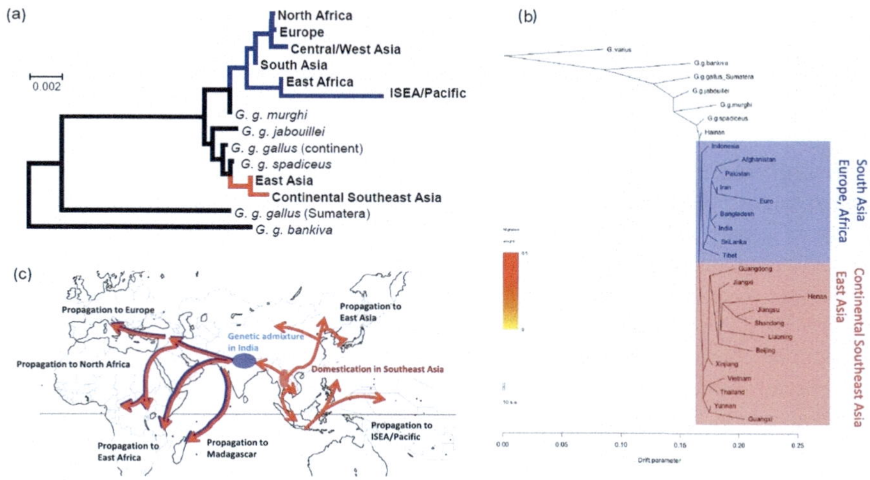

Fig. 6.1 Domestication and global propagation of chicken as inferred from the phylogeographic analyses. (**a**) Phylogenetic tree among populations based on the mitochondrial D-loop sequence (modified from Yonezawa and Sasaki 2016). Continental Southeast/East Asian chickens forms monophyletic group (colored in red) derived from *G. g. spadiceus*, and South Asian/European/African chickens forms monophyletic group (colored in blue) derived from *G. g. murghi*. This phylogenetic relationship suggests chickens were domesticated independently in continental Southeast Asia and South Asia at least from the maternal lineage. (**b**) Phylogenetic tree among populations based on the nuclear genome data (reanalysis of data from Wang et al. 2020). *G. g. spadiceus* and all domestic chicken populations form monophyletic group, suggesting the single domestication origin of chicken and that occurred in the continental Southeast Asia. 856 domestic chickens/junglefowls with 125,724 SNPs after the quality control and the linkage-disequilibrium pruning were used for the phylogenetic inference by using the TreeMix (Pickrell and Pritchard 2012) without assuming the migration events. (**c**) Evolutionary scenario of chicken domestication and global propagations. Chickens were initially domesticated in the continental Southeast Asia from the *spadiceus* subspecies, and propagated northward and reached East Asia, southward (eastward?) and reached ISEA/Pacific. One lineage propagated westward, reaching South Asia, where it interbred with the *murghi* subspecies. From this lineage, some lines spread further westward through West Asia to North Africa and Europe, while others migrated southward to East Africa and Madagascar. The results from Yonezawa et al. (2022, 2024) were also considered for this scenario. The map on the panel C was created from a white map provided by 3kaku-K (https://www.freemap.jp/) according to its terms of use

the *spadiceus* subspecies (Fig. 6.1b). This indicates a single domestication origin for the domestic chicken. Similar to the results of Lawal et al. (2020), Wang et al. (2020) indicated that domestic chickens have received genes introgressed from other subspecies, especially *murghi* during their spread to areas west of South Asia. Domestic chicken populations west of South Asia comprise monophyletic groups. This indicates that domestic chickens spread from South Asia to western Eurasia and is consistent with the results of the mitochondrial analyses. In contrast to the results of the mitochondrial analyses, domestic chicken populations west of South Asia are not derived from South Asian red junglefowl but instead from Southeast Asian domestic chickens.

Figure 6.1c shows a schematic diagram of chicken domestication and the global spread of chickens on the basis of these findings, including differences in the results of mitochondrial and nuclear analyses. Chickens were domesticated in continental Southeast Asia and spread widely throughout this region. Later, some lines spread northward into East Asia and southward (or eastward) into ISEA, whereas others spread westward into South Asia. This westward lineage is thought to have interbred with the *murghi* subspecies in South Asia. The genetic contribution from female *murghi* appears to have been highly significant in this hybridization event, given that the mitochondrial data indicate that populations west of South Asia were replaced by *murghi* lineages. This group spread further westward through West Asia to North Africa and Europe. A lineage from South Asia probably spread through the southern Arabian Peninsula to East Africa, and a group that spread directly across the Indian Ocean to Madagascar also emerged (Yonezawa et al. 2024).

6.4 Japanese Native Chickens in East/Southeast Asia

Next, I examine the evolutionary relationships of Japanese native chickens with Asian native chickens. Given that the results of the mitochondrial analysis indicate that domestic chickens in continental Southeast Asia and East Asia are derived from the subspecies *spadiceus* (Fig. 6.1a), I focus here on the evolutionary relationships of Japanese native chickens within continental Southeast Asia and East Asia populations. The phylogenetic network of domestic chickens in these regions on the basis of mitochondrial D-loop sequences by Yonezawa et al. (2020, submitted) is shown in Fig. 6.2. Here, each operational taxonomic unit (OTU) comprised native chickens derived from a single country; in China however, each OTU comprised native chickens derived from a single province. The phylogenetic network reflects a geographic cline from Southeast Asia through South and Middle China to North China (Central Plain). Populations of the historical peripheral regions in the Sinosphere (Chinese cultural sphere) are derived from North China and included Taiwan, Tibet, Xinjiang, Inner Mongolia, Northeast China, the Korean Peninsula, and Japan. This indicates that chickens were domesticated in Southeast Asia and then spread northward, throughout China, and eventually to surrounding areas. Japanese native chickens are closely related to chickens of North China and the peripheral regions of the Sinosphere, especially to those of the Korean Peninsula. These findings are consistent with zooarchaeological findings (Eda 2016, 2018).

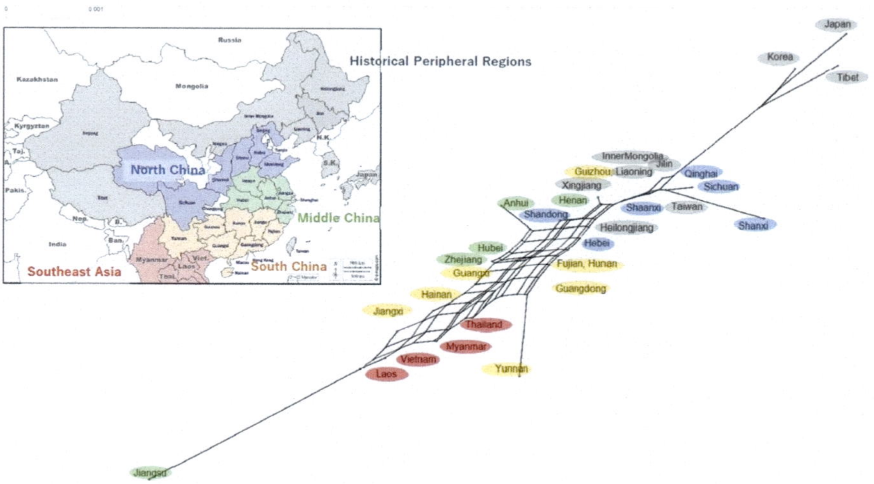

Fig. 6.2 Phylogenetic network of continental Southeast Asian and East Asian native chickens. The phylogenetic network was constructed based on the mitochondrial D-loop sequence (data from Yonezawa et al. (2020) and unpublished Okinawan native chickens). Continental Southeast Asia (red), South China (yellow), Middle China (green), North China (blue), and the historical peripheral regions of Sinosphere (black) are shown on the map in the upper left corner, and chickens in these regions are continuously arranged from south to north according to the geographical gradient in the phylogenetic network. Each population is color-coded according to its geographic distribution. Overall, Japanese chickens are found to be most closely related to native Korean chickens. The map on the upper left was created from a white map provided by the d-maps (https://d-maps.com/) according to its terms of use

6.5 History of Japanese Native Chickens According to Oana (1951)

Japanese native chickens are diverse; it is thus unlikely that all breeds were introduced to Japan via the Korean Peninsula. Therefore, I first introduce the currently accepted hypothesis by Oana (1951) (Fig. 6.3). Although proposed more than 70 years ago, this hypothesis is convincing because it was derived from a close examination of morphological features and an extensive literature review. The hypothesis is simple and provides an important starting point for understanding the overall history of Japanese native chickens. Oana (1951) first posits that Japanese chickens were derived from three main lineages that were introduced from different regions at different times. The first lineage is "*Jidori.*" *Jidori* is not a breed name but a generic term for unimproved breeds that retain ancestral characteristics. According to Oana (1951), *Jidori* was derived from the Korean Peninsula or China. The second lineage is "*Shokoku.*" *Shokoku* is an ornamental breed with long, graceful tail feathers and is thought to have been derived from China. *Shokoku* is thought to have been introduced to Japan during the Heian period (794–1185), given that a caricature of a chicken with the characteristics of *Shokoku* is depicted in "*Chōjū-jinbutsu-giga* (animal-person caricatures)"

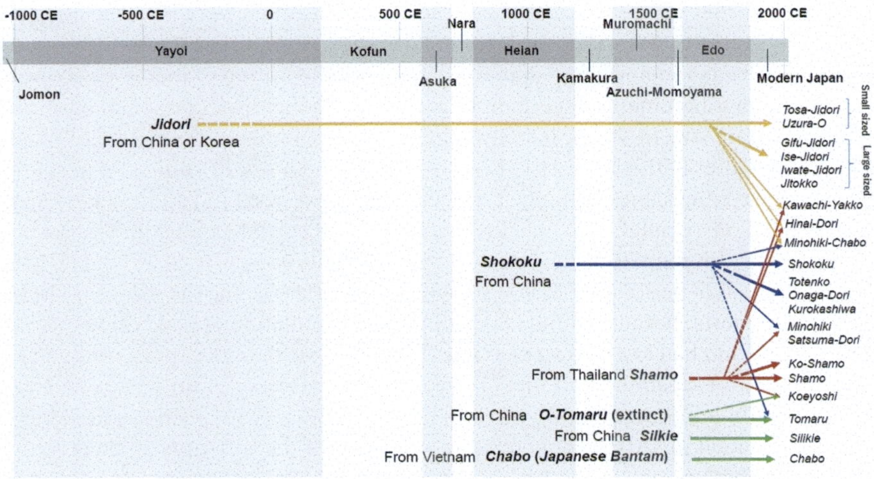

Fig. 6.3 Oana (1951)'s theories on the History of the Japanese native chicken. This figure was prepared based on the hypothesis of Oana (1951) with reference to Sato (2011) and Oka et al. (2007). The Japanese native chicken was based on *Jidori* (yellow), which was introduced from China or the Korean Peninsula during the Yayoi period, *Shokoku* (blue), which was introduced from China during the Heian to Kamakura periods, and *Shamo* (red), which was introduced from Thailand during the Edo period. The present variety of Japanese native chickens was established through direct breeding or crossbreeding among these lineages. During the Edo period, new lineages of chickens such as *Chabo* (Japanese bantam), *O-Tomaru*, and *Ukokkei* (Silkie) were also propagated from East and Southeast Asia (green)

created between the twelfth and thirteenth centuries (late Heian–early Kamakura period), and the name *Shokoku* appears in the diary of the poet Fujiwara no Sadaie (1162–1241) "*Meigetsu-ki.*" The third lineage is "*Shamo.*" *Shamo* is a large cockfighting breed and is thought to have been introduced from Thailand during the Edo period (1603–1868). The name "*Shamo*"is thought to be derived from "Siam," the old name for Thailand. During the national isolation policy in the Edo period, a variety of Japanese native chicken breeds were thought to be derived from these lines, either through direct breeding or crossbreeding among lines. It is also thought that "*Chabo* (Japanese bantam)," "*Silkie,*" and "*O-Tomaru*" were introduced from Vietnam and China between the Azuchi-Momoyama period (1573–1603) and the Edo period (see also Sato 2011; Imamura et al. 2021).

6.6 Evolutionary Relationships of Japanese Native Chickens with East/Southeast Asian Native Chickens

I discuss the history of Japanese native chickens from a molecular genetic perspective in light of Oana (1951)'s hypothesis. After dividing Japanese native chickens into different breeds, Yonezawa et al. (2020, submitted) examined their phylogenetic relationships with Southeast/East Asian native chickens on the basis of mitochondrial D-loop sequences (Fig. 6.4). Similar to the results of the analysis in which Japanese native chickens comprised a single OTU (Fig. 6.2), the results showed that each Japanese native chicken breed was genetically related to native chickens from Middle-North China and the peripheral regions. Japanese native chickens did not comprise a monophyletic group but showed a branching pattern that was repeatedly derived from Middle-North China and the peripheral regions. Japanese native chickens were divided into five major groups. The first four groups include *Tosa-Jidori*, *Ise-Jidori*, *Gifu-Jidori*, and *Aizu-Jidori*; these groups comprise almost exclusively Japanese native chickens. In contrast, the fifth group is scattered throughout the continental population in Middle-North China and the peripheral regions. The first four groups include breeds other than *Jidori* that are thought to have been derived from *Shokoku* (e.g., *Onaga-dori*, *Kurokashiwa*) or *Shamo* (e.g., *Koshamo*), or from crossbreeding between these lineages (e.g., *Satsuma-dori*) according to Oana (1951)'s hypothesis. In contrast, the fifth group primarily includes breeds thought to have been introduced directly from the continent after the Heian period (e.g., *Shokoku*, *Shamo,* and *Chabo*).

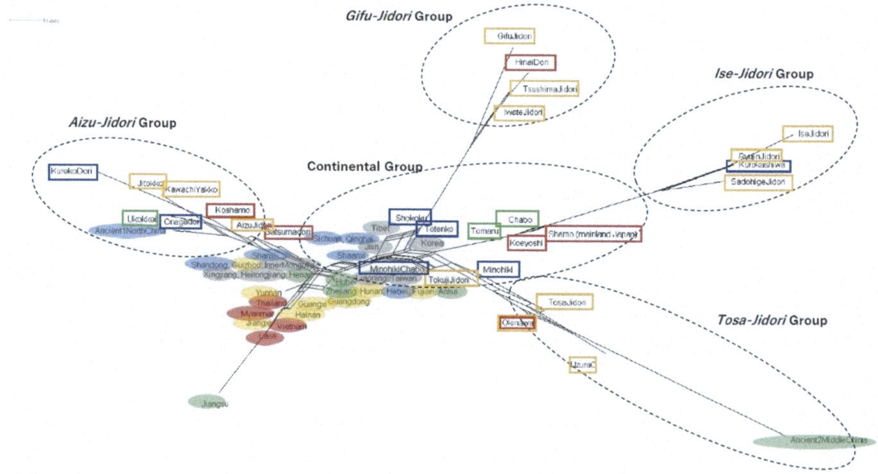

Fig. 6.4 Phylogenetic positions of Japanese native chicken breeds within continental Southeast Asian and East Asian native chickens. The data in Fig. 6.2 were reanalyzed after dividing the Japanese native chicken into breed levels. Color coding for each continental OTU is the same as in Fig. 6.2. The name of each breed of Japanese native chicken is enclosed in a square border, with the *Jidori*-related breeds in yellow, the *Shokoku*-related group in blue, the *Shamo*-related group in red, and the other groups in green. All Japanese native chicken breeds were placed in the northern East Asian group (Middle and North China as well as the peripheral regions of the Sinosphere). Japanese native chickens can be divided into distinct four lineages (the *Tosa-Jido*ri Group, the *Ise-Jidori* Group, the *Gifu-Jidori* Group, and the *Aizu-Jidori* Group) as well as the breeds (e.g., *Shokoku*, *Shamo*) highly intermingled with the continental populations. "Okinawa" includes three Okinawa native chicken breeds (*Ryukyu-Ko-Jidori*, *Chahn*, and *Taucii* (Okinawan variety of *Shamo*))

Although the history of Japanese native chickens from this phylogenetic network is difficult to interpret, the following conclusions can be made assuming that the first four lineages reflect the propagation of the ancestors of *Tosa-Jidori, Ise-Jidori, Gifu-Jidori*, and *Aizu-Jidori*. The ancestors of the *Jidori* lineages were derived from different regions of northern East Asia and were repeatedly introduced to Japan. As mentioned above, chickens in the Yayoi period had a marked sex ratio bias with a high preponderance of males, and it was thus difficult to maintain chickens in Japan for a long period at that time, and chickens might have been introduced from outside the country many times (Eda 2016). Our hypothesis is consistent with these archaeological findings.

The fifth group (*Shokoku, Shamo,* and *Chabo*) is nested within the continental population of northern East Asia. The haplotype composition of the fifth group is similar to that of the continental population, and the placement of *Shokoku*, which is thought to have spread at a more recent date, is consistent with Oana (1951)'s hypothesis. However, the fact that *Shamo* and *Chabo*, which are thought to have been derived from Thailand and from Vietnam, respectively, are related to northern East Asian populations, rather than Southeast Asian populations, provides important insights into the origin of these breeds. *Totenko*, which is thought to be derived from *Shokoku*, is located close to *Shokoku* in the mitochondrial phylogenetic network, and *Onaga-dori* and *Kurokashiwa*, which are also thought to be derived from *Shokoku*, comprise the *Aizu-Jidori* group. Similarly, *Koeyoshi*, which are thought to be derived from *Shamo*, is located close to *Shamo* in this phylogenetic network, and *Ko-Shamo*, which is also thought to be derived from *Shamo*, belonged to the *Aizu-Jidori* group. These findings suggest that some breeds thought to be derived from *Shokoku* or *Shamo* were established via hybridization with *Jidori* breeds.

6.7 Mitogenomic Analysis Provides a New Perspective on the Origin and History of Japanese Native Chickens

As mentioned above, weak phylogeographic structure was observed in domestic chickens in the mitochondrial analysis, and this increases the difficulty of inferring how domestic chickens have propagated via phylogenetic relationships among individuals. However, some mitochondrial haplogroups (or sub-haplogroups) are highly localized. These (sub-)haplogroups often show strong phylogeographic structure. Because these comprise a small portion of the total diversity of chickens, information on these (sub-)haplogroups alone does not provide comprehensive insights into the chicken propagation process, but they do provide important insights into the history of chickens.

I now focus on Sub-haplogroup C1, Haplogroup D, and Haplogroup H, which are present in Japanese native chickens. Haplogroup D is a rare haplogroup worldwide, but it is common in Pacific/ISEA chickens (Miao et al. 2013). Since

Haplogroup D occurs at a high frequency in some Japanese native chicken breeds (e.g., *Tosa-Jidori*), Pacific/ISEA chickens might have genetically contributed to Japanese native chickens (Oka et al. 2007). *Tosa-Jidori* is a small breed that retains many ancestral traits, and it is thought to have retained the characteristics of the first population introduced to Japan (Oana 1951). It is thus an important breed for understanding the history of Japanese native chickens. The fact that haplogroup D is observed at a relatively high frequency in *Tosa-Jidori* suggests that this haplogroup might be present in the population that gave rise to Japanese native chickens. Furthermore, haplogroup D is observed at a particularly high frequency (approximately 70%) in native chickens of Okinawa, especially in the southwestern portion of the Japanese Archipelago (Yonezawa et al. submitted). Given the proximity of Okinawa to the ISEA/Pacific region, chickens from Okinawa are critical for understanding the geographic origin of haplogroup D in Japanese native chickens.

I sequenced the mitochondrial genome of Japanese native chickens with Haplogroup D and inferred a phylogenetic tree (Yonezawa et al. 2022, submitted). Haplogroup D comprised Sub-haplogroup D2 in both the mainland and Okinawan populations (Fig. 6.5). Excluding Japan, Sub-haplogroup D2 has so far been reported only in Xinjiang (Miao et al. 2013; Huang et al. 2018). In contrast, both ISEA/Pacific native chickens belonged to Sub-haplogroup D1. Sub-haplogroup D3 is widely distributed from Middle China through Southeast Asia to South Asia (Miao et al. 2013, Huang et al. 2018). These phylogenetic relationships indicate that Haplogroup D, which is present in Japanese native chickens, spread from northern East Asia and possibly through mainland Japan to Okinawa.

Pacific native chickens are derived from ISEA, especially the Philippine population. The phylogenetic relationships among Pacific native chickens are highly consistent with the migration patterns of Austronesians (Yonezawa et al. 2022). Austronesians moved to Polynesia via Melanesia and eventually propagated to the Hawaiian Islands, the Marquesas Islands, and Easter Island (Fig. 6.5). This strongly suggests that the Austronesians had chickens as they spread throughout the Pacific. In addition, unlike chickens from other regions, strong phylogeographic structure was observed among native Pacific chickens. Austronesians are thought to have engaged in trade, but chickens may not have moved much after the Austronesians settled in these areas. In any case, this strong phylogeographic structure is useful for divergence time estimation.

Establishing calibration points on internal nodes of phylogenetic trees is extremely difficult when estimating evolutionary divergence times according to intraspecific variation. This stems from the difficulty of using paleontological (e.g., fossils) and archaeological (e.g., zooarchaeological remains) evidence to objectively assess where calibration points can be established on phylogenetic trees. Ancient DNA information can dramatically improve the accuracy of time estimation because it allows calibration points to be placed on terminal nodes rather than internal nodes. However, no ancient mitochondrial genome data are available for chickens. Therefore, the timing of chicken domestication is often estimated on the basis of mutation rates from Alexander et al. (2015)'s 50-generation pedigree analysis. However, the use of this mutation rate information to estimate evolutionary

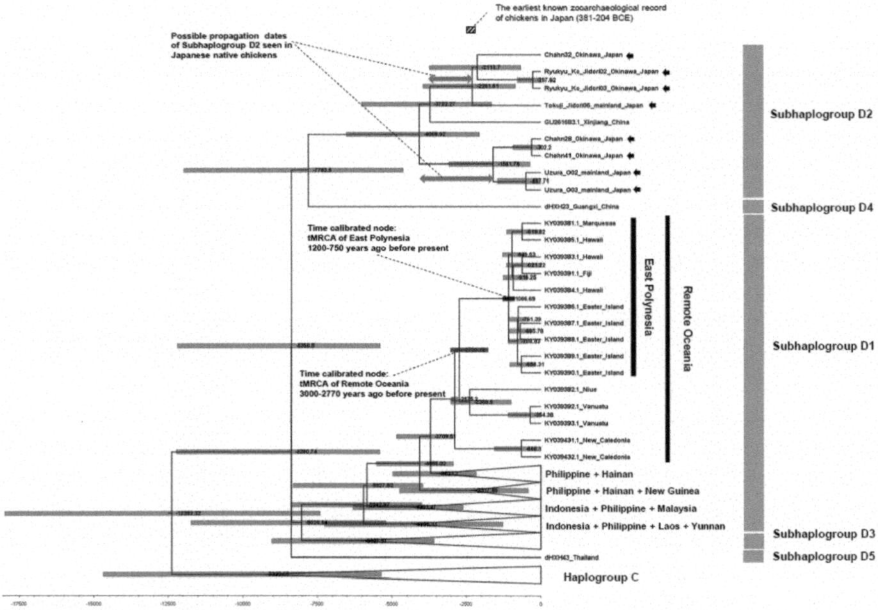

Fig. 6.5 Bayesian time tree of Haplogroup D based on the mitochondrial genome analysis. The branch lengths are proportional to the evolutionary time scale and nodal numbers indicate the coalescent times in years before present. The nodal bars indicate the 95% highest posterior densities of the coalescent times. The tMRCA of East Polynesian clade and that of the Remote Oceanian clade within the Subhaplogroup D1 were used for the time calibration points. All Japanese native chickens (indicated by the black arrows) in this haplogroup belong to Subhaplogroup D2. The gray colored double allows indicates the possible propagation dates of Japanese native chickens in this subhaplogroup. The shaded thick line indicates the age range (381–204 BC) of the oldest zooarchaeological record of chickens in Japan, as reported by Eda et al. (2023). Subhaplogroup D3 and several subclades within Subhaplogroup D1 is shown compressed. Haplogroup C is also shown compressed in this figure and its details are shown in Fig. 6.6

divergence times can result in extremely young age estimates compared with estimates on the basis of archaeological evidence (Huang et al. 2018). This may be because the molecular evolutionary rates on longer time scales is much slower than this mutation rate, since most mutations that occur within 50 generations are harmful and are quickly swept out.

The robust phylogeographic structure of Pacific native chickens can aid estimates of the evolutionary timing of chicken domestication. Recently, Peters et al. (2022) conducted a detailed analysis of zooarchaeological evidence of chickens from around the world and estimated the timing of their global spread. Zooarchaeological evidence of Pacific native chickens could be easily linked to phylogenetic clades because little migration occurred between regions once lineages settled. Thus, this zooarchaeological evidence can be used to establish calibration points and infer the divergence period (Yonezawa et al. submitted).

Assuming that chickens spread to the Japanese archipelago after they diverged from the most closely related lineages on the continent and before the time of the most recent common ancestor (tMRCA) in a clade that includes only Japanese individuals, the two sub-haplogroup D2 lineages of Japanese chickens emerged 4100–1600 and 3700–2300 years ago (Fig. 6.5). Assuming that these two lineages arrived in the Japanese archipelago at the same time, they are estimated to have emerged 3700–2300 years ago. This corresponds to the Late Jomon to Yayoi periods and is consistent with archaeological data indicating that chickens spread during the Yayoi period (Eda et al. 2023).

Next, I focus on Sub-haplogroup C1, which occurs at high frequency in East Asia but rarely in other regions (Miao et al. 2013). Huang et al. (2018) suggest that Haplogroup C might have originated in Southeast Asia, but Sub-haplogroup C1 might have spread from the Huang-Huai River Basin throughout East Asia. In addition, Sub-haplogroup C1 is particularly abundant in cockfighting breeds, suggesting that the spread of cockfighting culture might have affected the spread of Sub-haplogroup C1 (Huang et al. 2018).

The draft mitochondrial genome sequences of *Shamo* from mainland Japan (Aomori and Niigata Prefectures) and Okinawa were obtained and aligned with sequences from previous studies to infer a phylogenetic tree (unpublished). Individuals whose mitochondrial genomes were determined were randomly sampled with no prior information on mitochondrial haplotypes, but all individuals (five individuals) from the mainland and two of five individuals from Okinawa belonged to Sub-haplogroup C1. Of the seven *Shamo* individuals belonging to sub-haplogroup C1, four and two formed small, unique clades (*Shamo*-lineage 1 and *Shamo*-lineage 2), respectively (Fig. 6.6). Both of these clades contain both mainland and Okinawan *Shamo*. The place of origin could not be accurately estimated for the clade comprising four *Shamo* (*Shamo*-lineage 1) individuals. This is because multiple clades, including the *Shamo*-lineage 1, are derived from sub-haplogroup C1 in a multifurcation. Another clade comprising two *Shamo* individuals (*Shamo*-lineage 2) is derived from the South China (Guangdong, Guangxi, and Guizhou) populations, suggesting that this region might represent one of the origins of the maternal *Shamo* lineages. Middle China is also an origin of the *Shamo* breed given that one *Shamo* individual (*Shamo*-lineage 3) is nested in a clade comprising individuals from Anhui, Henan, Hunan, and Jiangxi (Fig. 6.6).

Sub-haplogroup C1 occurs at high frequency in several Japanese native chicken breeds (e.g., *Ise-Jidori*) (Yonezawa et al. 2020). Currently, there is not enough mitochondrial genome data of Japanese native chickens to clarify the phylogeographic structure within Sub-haplogroup C1. Therefore, there is also a possibility that these three maternal *Shamo* lineages might have been originated from *Jidori*, and introgressed into the ancestral *Shamo* population through crossbreeding with *Jidori*. An analysis of the *Shamo* lineages, as well as other Japanese native chickens such as *Shokoku* and *Jidori* is needed to clarify this possibility. In a previous study (Yonezawa et al. 2022), we found that the mitochondrial genome of one *Jidori* (*Aizu-Jidori*) individual belonged to this sub-haplogroup. This *Jidori* mitochondrial

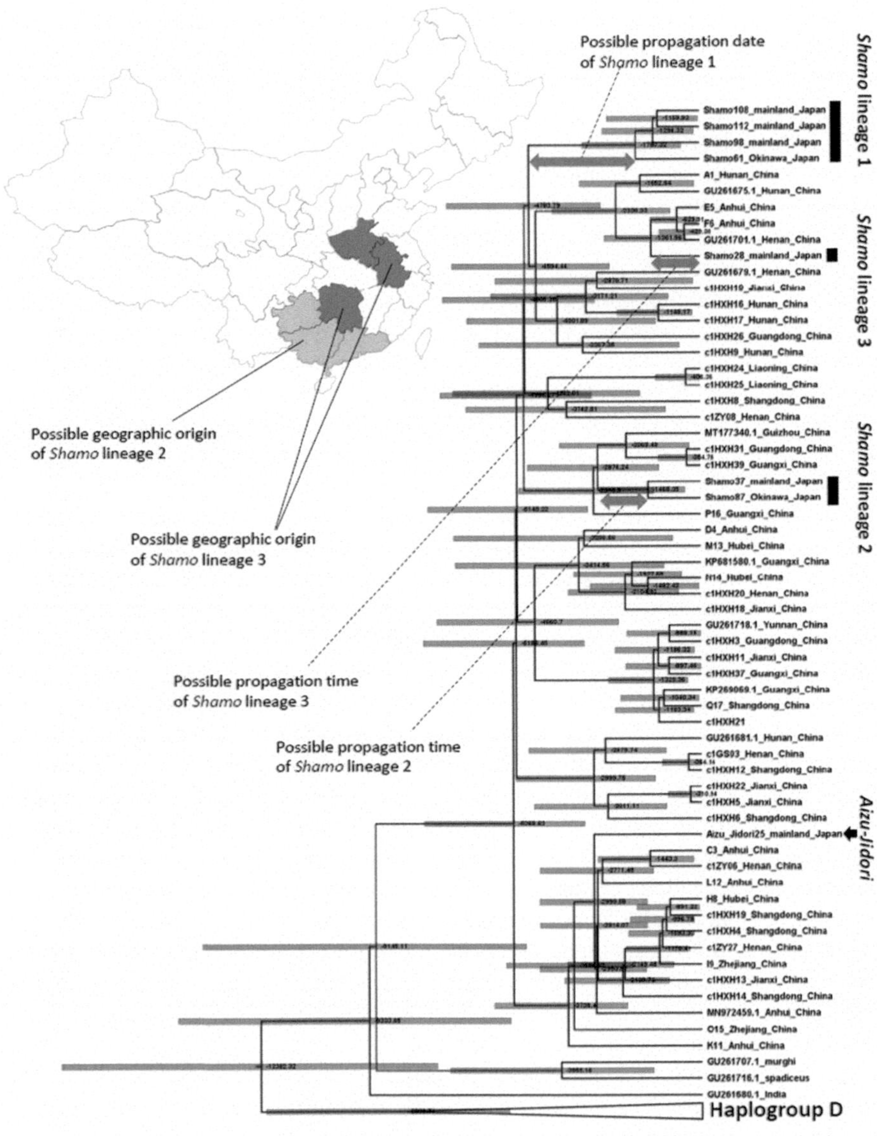

Fig. 6.6 Bayesian time tree of Haplogroup C based on the mitochondrial genome analysis. The branch lengths are proportional to the evolutionary time scale and nodal numbers indicate the coalescent times in years before present. The nodal bars indicate the 95% highest posterior densities of the coalescent times. Three *Shamo* lineages (thick black bars) and one *Aizu-Jidori* individual (black arrow) were labeled in this figure. The gray colored double allows indicates the possible propagation dates of these three *Shamo* lineages. The map in the upper left corner indicates the possible geographic origins of *Shamo*-lineage 2 (light gray) and *Shamo*-lineage 3 (dark gray). This map was created from a white map provided by 3kaku-K (https://www.freemap.jp/) according to its terms of use. Haplogroup D is shown compressed in this figure and its details are shown in Fig. 6.5

genome comprises a maternal lineage distinct from the three *Shamo* lineages (Fig. 6.6). Additional detailed analysis is needed.

The timing of the arrival of the *Shamo* ancestors to Japan merits investigation. Although we cannot rule out the possibility that the presence of Sub-haplogroup C1 in the *Shamo* lineage stems from genes introgressed from *Jidori* and other lineages, *Shamo* ancestors likely arrived in Japan after they diverged from the most closely related lineage in the continent and before the tMRCA in Japan if we assume that the three *Shamo* (maternal) lineages mentioned above are from the Chinese fighting cocks that were direct ancestors of the *Shamo*. *Shamo*-lineage 1 diverged from the most closely related continental lineages approximately 4700 years ago, and the tMRCA in Japan is approximately 1800 years ago; *Shamo*-lineage 2 diverged from the most closely related continental lineage approximately 3800 years ago, and the tMRCA in Japan is approximately 1500 years ago; and *Shamo*-lineage 3 diverged from the most closely related continental lineage approximately 1400 years ago (Fig. 6.6). Assuming that these three lineages arrived in Japan at the same time, they are estimated to have arrived in the first half of the first millennium. However, these arrival times might be overestimates. SHA61, which belongs to *Shamo*-lineage 1, has very long branches compared with those of the other individuals (data not shown). If this individual is excluded and the time tree is recalculated, *Shamo*-lineage 1, *Shamo*-lineage 2, and *Shamo*-lineage 3 were estimated to arrive in Japan between 4000 to 900 years ago, 2900 to 1400 years ago, and after 1400 years ago, respectively (data not shown). If these three lineages spread to Japan at the same time, they are estimated to have arrived in the second half of the first millennium.

Haplogroup H also merits special discussion. This is an extremely rare haplogroup. Other than the *Shamo* individual from Okinawa, this haplogroup has only been reported from Thailand and Xishuangbanna (Dai Autonomous Prefecture), Yunnan, China. This indicates that Haplogroup H is closely related to the Tai cultural sphere, which indicates that chickens from these areas have made genetic contributions to the Okinawan *Shamo* population. Oana (1951) states that *Shamo* chickens were introduced from Thailand. However, molecular genetic findings suggest that the main (matrilineal) ancestor of *Shamo* chickens is of East Asian origin, and that chickens from the Tai cultural sphere have contributed to the ancestral *Shamo* group (or at least the Okinawan population). This result provides partial support for the hypothesis of Oana (1951).

6.8 Revised History of Japanese Native Chickens

I reconsider the history of Japanese native chickens in light of the above findings and the hypothesis of Oana (1951). Below, I outline an evolutionary scenario for their origin and history according to the results of molecular genetic and archaeological studies (Fig. 6.7).

Phase I. Propagation of *Jidori* (Yayoi-Kofun Periods): *Jidori* is thought to have spread to the Japanese Archipelago multiple times from an area spanning from

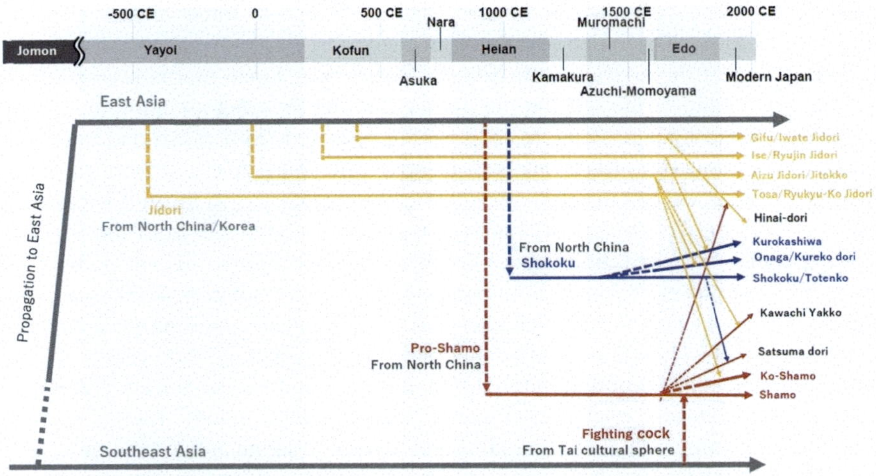

Fig. 6.7 The origin and history of the Japanese native chicken, with a reconsideration of Oana (1951)'s hypothesis, taking into account the results of phylogeographical analysis. Including *Shamo*, which has been considered to have originated in Southeast Asia, Japanese native chickens have their main origins in northern East Asia. *Jidori* breeds are thought to have propagated repeatedly from northern East Asia to the Japanese archipelago. *Shokoku* and *Pro-Shamo* (hypothetical ancestor of *Shamo*) are also thought to have been propagated from northern East Asia, but the *Onag-dori*, *Kurokashiw*, and *Ko-Shamo*, which are thought to have been derived from these breeds, are thought to have received significant genetic contributions from *Jidori*. On the other hand, some *Shamo* population, especially Okinawan population (*Taucii*), have mitochondrial Haplogroup H characteristic of Tai cultural sphere, suggesting that they have been genetically influenced by cockfighting breeds from these regions for the establishment of the extant *Shamo* population

Middle China to the Korean Peninsula. Zooarchaeological evidence suggests that *Jidori* chickens arrived in Japan from the Yayoi period or later (Eda et al. 2023). Rice cultivation was introduced from the Asian continent (China and the Korean Peninsula) during this period. Therefore, it is safe to assume that chickens were introduced along with the rice culture. The Yayoi period in Japan marked the transition from a hunter-gatherer society to an agricultural society. This was a cultural transition that resulted in genetic changes in the Japanese people, given that it was caused by the arrival of farmers from the Asian continent. Chickens were probably brought to the Japanese Archipelago by these continental agriculturalists. Given that the large-scale arrival of farmers is thought to have continued until the Kofun period (third–seventh Century), chickens were repeatedly introduced to the Japanese archipelago between the Yayoi and Kofun periods (Eda 2018), and the substratum of the Japanese native chicken was probably formed. Chickens from this period probably played a greater role as prestige goods than as a source of animal protein, given the extremely small number of chicken remains excavated from archaeological sites, the absence of cut marks on these chickens, and the fact that, as mentioned above, the sex ratio bias was significant, which increased the difficulty of maintaining the chicken population within the country (Eda 2023).

Jidori might have been introduced to Japan at least four times. That is, the ancestors of the *Tosa-Jidori*, *Ise-Jidori*, *Gifu-Jidori*, and *Aizu-Jidori* groups might have arrived in Japan via four independent events. Determining when and from what regions the ancestors of these four groups were derived remains a challenge, but ancient DNA data from chickens excavated from Chinese archaeological sites (Xiang et al. 2014) might provide important insights. The *Aizu-Jidori* group is similar in genetic composition to the North Chinese (Shandong and Hebei) population from over 4000 years ago; it is also closely related to the extant Shanxi (North China) population (Fig. 6.4). The *Tosa-Jidori* group is related to a population in Middle China from approximately 2300 to 3000 years ago (Fig. 6.4). Sub-haplogroup D2 has been detected in the *Tosa-Jidori* group. Whole mitochondrial genomes are needed to classify haplogroup D into sub-haplogroups. Although the ancient DNA data from Xiang et al. (2014) comprise partial D-loop sequences and thus cannot be used for sub-haplogroup classification, these data indicate a temporary increase in the frequency of Haplogroup D in Middle China approximately 2300–3000 years ago (corresponding to the Yayoi period in Japan), and the ancestors of the *Tosa-Jidori* group might have been introduced to Japan during this period.

The repeated introduction of chickens to the Japanese Archipelago might have increased the genetic diversity of Japanese native chickens and thus led to the accumulation of diverse mitochondrial haplogroups in the gene pool. The origin and history of Okinawa native chickens merit examination. The extremely high frequency of Haplogroup D (approximately 70%) in Okinawa native chickens might stem from its spread from mainland Japan to Okinawa before diverse haplogroups appeared in the gene pool of chickens in mainland Japan. The history of Okinawa was independent of that of mainland Japan until the nineteenth century, and the transition from a hunter-gatherer society to an agricultural society is believed to have occurred after the tenth century (Takamiya et al. 2016). Therefore, the introduction of chickens to Okinawa is thought to have occurred after the tenth century, but records from the Yi Dynasty of Korea indicate that chickens had spread throughout Okinawa by the fifteenth century at the latest (Yonezawa et al. submitted).

Recently, an unimproved chicken population was discovered in Okinawa (Takada 2020). It is likely that an extinct population kept in farmer's yards as "*Paa-Tui*" (*Paa* means "red" and *Tui* means "birds" (or "chickens") in the local dialect of the Okinawan language) in Okinawa, survived in a feral state. Takada (2020) describes this population as *Ryukyu-Ko-Jidori* (*Ryukyu* is the old name of Okinawa, and *Ko* means "ancient" in Japanese). Similar to *Tosa-Jidori*, the *Ryukyu-Ko-Jidori* population retains the small stature and ancestral traits reminiscent of red junglefowl. Given that Sub-haplogroup D2, which is rare in East Asia, occurs at a high frequency in the *Ryukyu-Ko-Jidori* population, these breeds are important for clarifying the history of native chickens in Japan and East Asia. In addition, *Tosa-Jidori*, *Ryukyu-Ko-Jidori*, and other *Jidori* chickens from various regions of Japan are important for clarifying the origin of Japanese native chickens. Additional detailed analyses are needed.

Phase II. Propagation of *Shokoku* and *Pro-Shamo* (Heian-Kamakura Periods): The foundation of Japan as an ancient nation was established from the

Yayoi period to the Kofun period, and this was followed by the establishment of the centralized nation (Yamato kingdom). During this process, the role of *Jidori* might have gradually changed. In the middle Kofun period, chicken remains with cut marks have been identified, suggesting that chickens began to be used as a source of food (Eda 2023). During the Asuka period (592–710), domestic rule by the Japanese Imperial Court stabilized, and in the following Nara (710–794) and Heian (794–1185) periods, Japan experienced many cultural exchanges through trade with China and the Korean Peninsula. Descriptions of *Shokoku* and fighting cocks began to appear in paintings and documents established during the Heian and Kamakura periods (1185–1333). A new culture of chickens, including cockfighting and ornamental breeds, was introduced from China. The fighting cocks introduced from China are thought to be the ancestors of the *Shamo* breed. I refer to these fighting cocks as *Pro-Shamo*. Various breeds were derived from *Shokoku* and *Pro-Shamo*, and these were likely crossbred with *Jidori* during this period. *Pro-Shamo* is thought to have spread to Okinawa.

Phase III. Establishment of diverse Japanese native chickens: Even after the Kofun period and until the beginning of the Edo period (1603–1868), chicken remains were not commonly excavated from archaeological sites (Eda 2023). However, chicken breeds began to diversify by the Azuchi-Momoyama period (1573–1603). Several chicken remains from this period had cut marks, indicating that chickens were important food sources. Chickens of this period were also increasing in size, and a discontinuous distribution of body size in both sexes was observed, which indicates the diversification of breeds (Eda 2021, 2022). In addition, various breeds were imported from China and Southeast Asia, including *Silkie*, *Chabo* (Japanese bantam), and the extinct *O-Toumaru* (Sato 2011).

During this process, cockfighting breeds were introduced from the Tai cultural sphere, probably via Okinawa. At that time, Okinawa comprised an independent nation known as the Ryukyu Kingdom (1429–1879) and played an important role as a trading post between Japan, China, and Southeast Asian countries. The fact that Haplogroup H is reported only from Okinawa and the Tai cultural sphere indicates that trade was conducted in both spheres, and this provides insights into the history of Okinawa *Shamo*. The lack of Haplogroup H from mainland Japan means that only male Okinawa *Shamo*, which was derived from crossbreeding between Thai fighting cocks and Okinawan *Pro-Shamo*, was introduced to mainland Japan.

Jidori served as the founding population and facilitated the establishment of the Japanese native chicken population via the breeding of *Shokoku* and *Pro-Shamo*. *Silkie*, *Chabo*, and *O-Tomaru* also made genetic contributions to Japanese native chickens, and further breeding and crossbreeding between lineages during the policy of national isolation in the Edo period led to the formation of a variety of Japanese chickens.

Acknowledgments I wish to express my special thanks to Dr. Yoshio Yamamoto, Dr. Masahide Nishibori, Dr. Jiaqi Wu, Dr. Yoshiaki Nakamura, Dr. Masaoki Tsudzuki, Dr. Hideyuki Mannen, Dr. Keisuke Tanaka, Mr. Masaru Takada, Mr. Yasutaka Imamura, Mr. Satoru Fukagawa, Mr. Shogo Matsunaga, Dr. Masaharu Matsuzaki, Dr. Hiroshi Ogawa, Dr. Takeshi Sasaki, Dr. Hitomi Hongo,

Dr. Kazunobu Ikeya, Dr. Naoki Tsunekawa, Dr. Yoshihiro Hayashi, Dr. Hirohisa Kishino, Dr. Masami Hasegawa, Dr. Masaki Eda, Dr. Kohei Kudo, Dr. Hideki Endo, and HIH Crown Prince Fumihito Akishinonomiya, Ph. D for discussions on the origins and histories of domestic chickens and other native livestock. I also thank Dr. Chris Akcali (Edanz: https://jp.edanz.com/ac) for editing a draft of this manuscript.

Competing Interests The author has no relevant financial or non-financial interests to disclose.

References

Akishinonomiya F, Miyake T, Takada M et al (1996) Monophyletic origin and unique dispersal patterns of domestic fowls. Proc Natl Acad Sci USA 93:6792–6795. https://doi.org/10.1073/pnas.93.13.6792

Alexander M, Ho SYW, Molak M et al (2015) Mitogenomic analysis of a 50-generation chicken pedigree reveals a rapid rate of mitochondrial evolution and evidence for paternal mtDNA inheritance. Biol Lett 11:20150561. https://doi.org/10.1098/rsbl.2015.0561

Darwin C (1868) The variation of animals and plants under domestication, 1st edn. John Murray, London

Eda M (2016) Chickens in the Yayoi period: in the context of osteological microevolution under domestication. Zoo-Archaeology 33:49. (in Japanese)

Eda M (2018) Yayoi Jidai No Niwatori, Saikou. Kikan Koukogaku 144:43–46. (in Japanese)

Eda M (2021) Size of domestic chickens in medieval Japan, based on chicken remains from the Otomo-Funai site, Oita. Bull Lab Archaeol Hokkaido Univ 1:41–50. (in Japanese)

Eda M (2022) Shape of domestic chickens in the medieval Japan, based on chicken remains from the Otomo-Funai site, Oita. Bull Lab Archaeol Hokkaido Univ 2:43–53. (in Japanese)

Eda M (2023) The history of chickens in Japanese archipelago based on archaeological materials. Biostory 40:54–59. (in Japanese)

Eda M, Izumi H, Yoneda M et al (2023) The earliest evidence of domestic chickens in the Japanese archipelago. Front Earth Sci 11:1104535. https://doi.org/10.3389/feart.2023.1104535

Eriksson J, Larson G, Gunnarsson U et al (2008) Identification of the yellow skin gene reveals a hybrid origin of the domestic chicken. PLoS Genet 4:e1000010. https://doi.org/10.1371/journal.pgen.1000010

Hata A, Takenouchi A, Kinoshita K et al (2020) Geographic origin and genetic characteristics of Japanese indigenous chickens inferred from mitochondrial D-loop region and microsatellite DNA markers. Animals 10:2074. https://doi.org/10.3390/ani10112074

Herrera MB, Kraitsek S, Alcalde JA et al (2020) European and Asian contribution to the genetic diversity of mainland South American chickens. Roy Soc Open Sci 7:191558. https://doi.org/10.1098/rsos.191558

Huang XH, Wu YJ, Miao YW et al (2018) Was chicken domesticated in northern China? New evidence from mitochondrial genomes. Sci Bull 63:743–746. https://doi.org/10.1016/j.scib.2017.12.004

Imamura Y, Tsudzuki M, Roszkowski S (2021) Japanese chickens: the living art of the world. Taurus Printing. Durban, South Africa

Kanginakudru S, Metta M, Jakati RD et al (2008) Genetic evidence from Indian red jungle fowl corroborates multiple domestication of modern day chicken. BMC Evol Biol 8:1–14. https://doi.org/10.1186/1471-2148-8-174

Lawal RA, Martin SH, Vanmechelen K et al (2020) The wild species genome ancestry of domestic chickens. BMC Biol 18:1–18

Lawler A (2014) Why did the chicken cross the world? The epic Saga of the bird that powers civilization. Atria, New York, NY, p 324. ISBN 978-1-4767-2989-3

Liu YP, Wu GS, Yao YG et al (2006) Multiple maternal origins of chickens: out of the Asian jungles. Mol Phylogenet Evol 38:12–19. https://doi.org/10.1016/j.ympev.2005.09.014

Miao YW, Peng MS, Wu GS et al (2013) Chicken domestication: an updated perspective based on mitochondrial genomes. Heredity 110:277–282. https://doi.org/10.1038/hdy.2012.83

Oana H (1951) The history of Japanese native chicken. Nihonkei-Kenkyusha, Tokyo. (in Japanese)

Oka T, Ino Y, Nomura K et al (2007) Analysis of mtDNA sequences shows Japanese native chickens have multiple origins. Anim Genet 38:287–293. https://doi.org/10.1111/j.1365-2052.2007.01604.x

Osman SAM, Sekino M, Kuwayama T et al (2006) Genetic variability and relationships of native Japanese chickens based on microsatellite DNA polymorphisms-focusing on the natural monuments of Japan. J Poult Sci 43:12–22. https://doi.org/10.2141/jpsa.43.12

Peters J, Lebrasseur O, Irving-Pease EK et al (2022) The biocultural origins and dispersal of domestic chickens. Proc Natl Acad Sci USA 119:e2121978119. https://doi.org/10.1073/pnas.2121978119

Pickrell JK, Pritchard JK (2012) Inference of population splits and mixtures from genome-wide allele frequency data. PLoS Genet 8:e1002967. https://doi.org/10.1371/journal.pgen.1002967

Sato S (2011) Origin and definition of Japanese native chickens, "Jidori". J Jpn Soc Poult Dis 47:1–11. (in Japanese)

Takada M (2020) *Paa-Tui* (Kashou: Ryukyu-Ko-Jidori) no Shoukai. Kachiku Shigen Kenkyukaiho 19:6–7. (in Japanese)

Takamiya H, Hudson MJ, Yonenobu H et al (2016) An extraordinary case in human history: prehistoric hunter-gatherer adaptation to the islands of the central Ryukyus (Amami and Okinawa archipelagos), Japan. Holocene 26:408–422. https://doi.org/10.1177/0959683615609752

Tsudzuki M (2003) Japanese native chickens. In: Chang HL, Huang YC (eds) The relationship between indigenous animals and humans in APEC region. The Chinese Society of Animal Science, Tainan, pp 91–116

Wang MS, Thakur M, Peng MS et al (2020) 863 genomes reveal the origin and domestication of chicken. Cell Res 30:824–825. https://doi.org/10.1038/s41422-020-0349-y

Xiang H, Gao J, Yu B et al (2014) Early Holocene chicken domestication in northern China. Proc Natl Acad Sci USA 111:17564–17569. https://doi.org/10.1073/pnas.1411882111

Yonezawa T, Sasaki T (2016) Idenshi ni Kizamareta Niwatori no Kita Michi. Kachiku Shigen Kenkyukaiho 15:4–13. (in Japanese)

Yonezawa T, Tsudzuki M, Imamura Y et al (2020) The origin and history of native Japanese chickens based on the mitochondrial DNA. Archivio per l'Antropologia e la Etnologia 150:67–88

Yonezawa T, Nishibori M, Yamamoto Y et al (2022) Complete mitochondrial genome analysis clarifies the enigmatic origin of Haplogroup D in Japanese native chickens. J Poult Sci 59:316–322. https://doi.org/10.2141/jpsa.0220027

Yonezawa T, Mannen H, Honma K et al (2024) Origin and spatial population structure of Malagasy native chickens based on mitochondrial DNA. Sci Rep 14:569. https://doi.org/10.1038/s41598-023-50708-x

Yonezawa T, Takada M, Tanaka K et al (submitted) Origin and diversity of Okinawa native chickens as inferred from mitochondrial DNA analysis

Chapter 7
Evaluating the Interesting Journey of Taro Plant Varieties to Japan Using Phylogenetic Analysis

Reiko Motohashi, Yuina Saitou, Natsuki Shinomura, Tatsuo Konishi, Hiroshi Fujii, Mitsuo Omura, and Naoki Osada

Abstract Taro (*Colocasia esculenta*) is thought to be one of the first crops cultivated by humans, and was an important crop cultivated before the domestication of rice. Previous research has hypothesized that taro strains with less astringent taste profiles were selected from wild taro by humans. In this study, we evaluate the route by which taro arrived in Japan (Yaponesia). We report the results of phylogenetic analyses of taro that combine polyploid data with chloroplast, simple sequence repeat, and single nucleotide polymorphism markers. Finally, we also discuss the relation between Yaponesian and taro.

Keywords Taro · *Colocasia* · First crop · Starch source · GRAS-Di

Abbreviations

BWA	Burrows-Wheeler aligner
CV	Cultiva

R. Motohashi (✉)
Research Institute of Green Science and Technology/Graduate School of Integrated Science and Technology, Shizuoka University, Shizuoka, Japan
e-mail: motohashi.reiko@shizuoka.ac.jp

Y. Saitou · N. Shinomura · M. Omura
Graduate School of Integrated Science and Technology, Shizuoka University, Shizuoka, Japan
e-mail: shinomura.natsuki.18@shizuoka.ac.jp; omura.mitsuo@shizuoka.ac.jp

T. Konishi
The Research Institute of Evolutionary Biology, Tokyo, Japan
e-mail: t2konishi@hi3.enjoy.ne.jp

H. Fujii
Department of Agriculture, Shizuoka University, Shizuoka, Japan
e-mail: fujii.hiroshi@shizuoka.ac.jp

N. Osada
Faculty of Information Science and Technology, Hokkaido University, Sapporo, Hokkaido, Japan
e-mail: nosada@ist.hokudai.ac.jp

N. Osada et al. (eds.), *Phylogeographic History of Plants and Animals Coexisting with Humans in Asia*, Evolutionary Studies,
https://doi.org/10.1007/978-981-97-6887-5_7

GRAS-Di	Random amplicon sequencing-direct
RAD-seq	Restriction-site associated DNA sequencing
RFLP	Restriction fragment length polymorphism
SNP	Single nucleotide polymorphism
SSR	Simple sequence repeat
WT	Wild Taro

7.1 Taro

"Satoimo" is the Japanese name for taro (*Colocasia esculenta*, Araceae), a widely distributed crop with a name borrowed from the Maori language. Taro is globally distributed throughout the tropical and temperate zones of Africa, Eurasia, Oceania, and America (Matthews 1991; Rao et al. 2010; Matthews and Ghanem 2021). All wild species of *Colocasia* except *C. esculenta* are native to the moist montane forests of Northeastern India, East Asia, and Southeast Asia. It appears to have evolved as a species adapted to wetlands, which partly explains how it spread—both as a cultivated and a wild plant—throughout the densely populated Asian landscape.

The genus *Colocasia* is thought to be related to the genera *Alocasia*, *Xanthosoma*, and *Cyrtosperma*. Similar taro species include giant taro (*Alocasia macrorrhizos*), swamp taro (*Cyrtosperma merkusii*), and arrowleaf elephant's ear (*Xanthosoma*

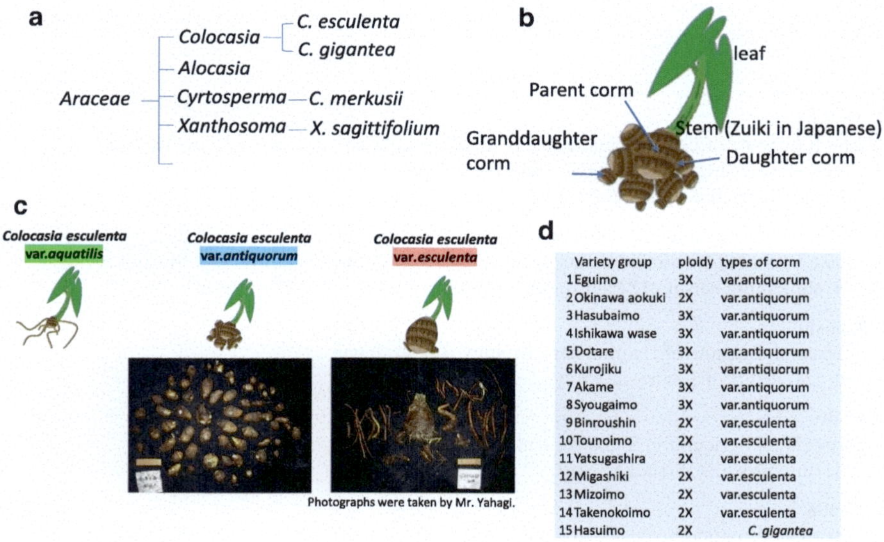

Fig. 7.1 Morphological characteristics of Japanese and non-Japanese taro varieties. (**a**) The organization of the family *Araceae*. (**b**) Characteristic of taro morphology. (**c**) Three types of corm morphology. (**d**) Groups of Japanese varieties reported by Kumazawa et al. (1956)

sagittifolium) (Fig. 7.1a). Taro comprises diploid (2n = 28) and triploid (3n = 42) species that produce offspring mainly via vegetative propagation. Morphologically, taro is highly plastic, and shows three morphotypes: var. *aquatilis* (which features long stolons), var. *antiquorum* (dasheen type, which features many secondary corms), and var. *esculenta* (eddo type, which features a large parental corm) (Plucknett 1983) (Fig. 7.1b, c). Many similar and homologous taro varieties exist, and in Japan the species was divided into 15 varietal groups by Kumazawa et al. (1956) (Fig. 7.1d).

Molecular biological research on taro has begun only recently. For example, a group from Hawaii reported the genome sequences of two varieties, including a diploid Samoan line known as "Niue" (GenBank: GCA_009445465.1, 2019), which is sensitive to taro Leaf Blight, and a relatively disease-resistant line called "Moi" (GenBank: GCA_013364625.1, 2020) (Helmkampf et al. 2018; Bellinger et al. 2020). Genome sequences determined using genotyping by sequencing and restriction-site associated DNA sequencing (RAD-seq) approaches have also been performed. Although the coverage of the sequenced genome was low, the authors identified single nucleotide polymorphism (SNP) markers spanning the entire genome. Complete genome sequences for individual taro lines were difficult to determine since the diploid line "Moi" had a genome size of 2.45 Gb due to its hemiploid assembly, while the triploid lines had a genome size of nearly 10 Gb (Bellinger et al. 2020). A high-quality whole genome sequence of the Chinese cultivar "Longxiangyu" (2n = 2x = 28, 2.4 Gb, GenBank: GCA_014218235.1, Accession: PRJNA587719), was also obtained (Yin et al. 2021). In this genome assembly, 88.43% of the sequence (2.1 Gb) was found to be repetitive, and we predicted 28,695 protein-coding genes. Most of the repetitive sequences were LTR-type retrotransposons. Furthermore, 44.71% of the genome was *gypsy*-type and 28.4% was *copia*-type, suggesting that *gypsy* was the major driving force in the evolution of the taro genome (Yin et al. 2021). However, to date most taro genomes sequenced were diploid lines. Since triploid lines are widely grown in Japan, additional genome information characterizing triploid taro is also required.

7.2 Ethnocultural Evidence of Ancient Taro in Japan

Taro (along with yams, sugarcane, and bananas, among other plants) are believed to have been cultivated by agrarian cultures present in Southeast Asia between 10,000 and 20,000 years ago. It is therefore considered to be an ancient crop (Rao et al. 2010). Although various theories exist regarding the origin of taro in India and China, many researchers infer that it originated in or near eastern India before spreading to the west and east via cultivation, possibly due to the abundance of wild taro lineages and the existence of favorable climatic conditions. The fact that taro fruits do not contain hard parts that resist rotting makes it difficult to preserve taro as a fossil and has made research accordingly difficult. Although no direct evidence has yet been found, charred taros have been excavated at archaeological sites, suggesting that taro came to Japan with the migration of the Yaponesians, the ancestors of the Jomon people, and was therefore cultivated in Japan prior to rice cultivation. In Japan, taro is found in many wilderness areas. One such cultivar is a taro variety present in the Kutsukake hot spring area in Nagano Prefecture. These are hypothesized to have survived since the Jomon Era. During this period the average temperature was 2 ~ 3 °C higher than the present, and ancient taro was transported to Japan, after which it was widely distributed in Japan from southern to northern parts of the country. However, during the cold period spanning from the late Jomon Era to the Yayoi Era, many taro cultivars became extinct, and those that survived may have done so by becoming wild in areas with favorable growing conditions, including hot springs.

Several wild taro cultivars have been reported as existing in medieval Japan, as reported by the legend of Kōbō Daishi's stone taro. This legend is as follows and follows Kūkai (July 27, 774–April 22, 835), posthumously called Kōbō Daishi, a Japanese Buddhist monk. Once an old woman was washing taros, and a shabbily dressed traveling monk (Kōbō -Daishi) passed by and asked her for some taros. The greedy old woman refused, saying that the taros were too hard or too bitter to eat. These taros then became as hard as stone or so bitter that it caused her tongue to bend.

According to the Kojiki, Nihonshoki, and Izumo no Kuni Fudoki, three important texts written in 712, 720, and 733, respectively, Tsukuyomi no Mikoto (Tsukuyomi), the god of the moon, was respected in ancient societies. The ancient people within these societies revered the moon via moon dances, which were associated with taro. In many cases, sumo wrestling is also related to celebration of the moon and fertility rites such as Jugoya (i.e., the night of the 15th day of the eighth lunar month). Moreover, this rite is held during the Zuiki (i.e., "stem of taro") Festival. Perhaps the origin of the Zuiki Festival and sumo wrestling can both be investigated further to generate ethnocultural evidence of the timing of the taro's arrival in Japan.

For the celebration of the 470-year-old Zuiki festival at Mikami Shrine (Fig. 7.2g), taros grown in specific five districts (i.e., the Torinji, Yamade, Maeda, Konakakoji, and Onakakoji districts) are used (Fig. 7.2). Importantly, taro is highly sensitive to cold and drought. The wild tubers around Kutsukake hot spring in Aoki Village are themselves grown in hot springs, so they may have survived due to warm water from hot springs during the cold period spanning from the late Jomon Era to the Yayoi Era. Similarly, the wild-grown taros present at the spring of Ryugenji Temple in Kofu City (Yamanashi Prefecture), Taro Valley spring (Izumo City, Shimane Prefecture), and Eguimo Street (Sekigane hot spring, Tottori Prefecture) have been regarded as descending from Kōbō Daishi's legendary stone taros.

In this study, we collected cultivated taros as well as wild taros from many sites in Asia and Oceania. We determined the sequence of the PS-ID (plastid subtype ID) region of the chloroplast genome to identify the origin from maternal genetics. Moreover, simple sequence repeat (SSR) marker analysis and Random Amplicon Sequencing-Direct (GRAS-Di) analysis were used for analyzing these taros. The genotyping was performed using these data. In addition, as an outgroup for

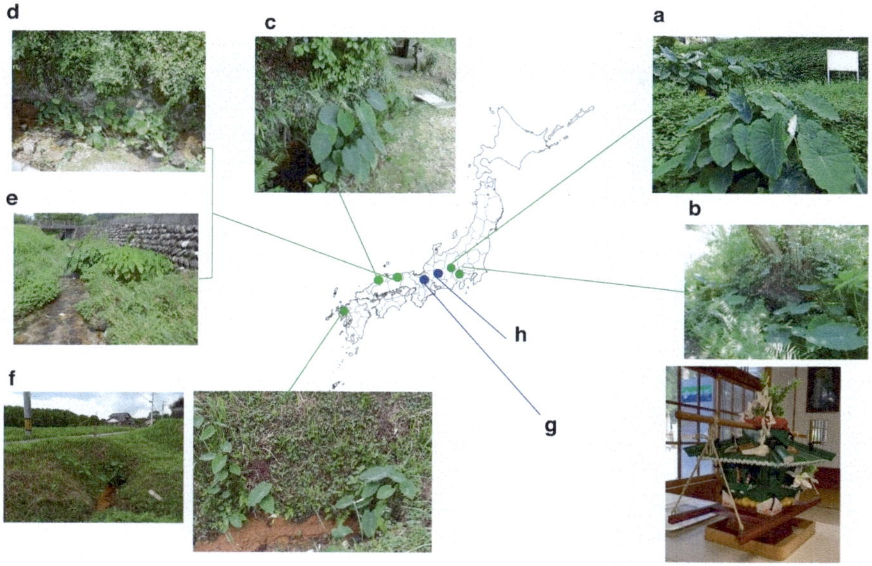

Fig. 7.2 Old Japanese taro varieties including wild-type (green spots) and cultivar (blue spots). (**a**) Kutsukake hot spring (geographic coordinates: 36°21′0′′N and 138°06′45.3′′E). (**b**) Ryugenji Temple (geographic coordinates: 35°41′49.2″N and 138°32′35.4″E). (**c**) Taro Valley (Oono-cyou, Matsue City, Shimane Prefecture). (**d**) Sekigane hot spring (Sekiganejyuku, Sekigane-cyou, Kurayoshi City, Totori Prefecture). (**e**) Sakura embankment (Tenjin river park, Ooaza tagiri, Yunashihama-cyou, Touhaku-gun, Totori Prefecture). (**f**) Tosu City (Takeshita et al. 1991). (**g**) Mikoshi made using taro stems for the Zuiki Festival at the Mikami Shrine (Mikami 838, Nosu City, Shiga Prefecture). (**h**) Nishikata taro (Kashimo, Nakatsugawa city, Gifu Prefecture)

phylogenetic analysis we used *C. gigantea*, whose petioles are used for food, and the American taro *Xanthosoma*, which is native to tropical regions of Central and South America (Fig. 7.1a).

7.3 Chloroplast Genome

First, we conducted a phylogenetic analysis using sequences from the PS-ID region of the chloroplast genome. We used a total of 99 samples of the "Eguimo," "Hasuba imo," "Dotare," "Taira-ooba," "Akame," "Celebes," "Kurojiku," "Taimo," "Tounoimo," "Takenokoimo," "Binroushin," and "Yatsugashira" taro varieties, as well as *C. gigantia* and *Xanthosoma*. These samples were collected from a total of 20 countries, including Japan, China, Taiwan, South Korea, India, Nepal, Pakistan, the Philippines, Malaysia, Indonesia, Thailand, Tonga, Hawaii (USA), Papua New Guinea, American Samoa, Costa Rica, Brazil, Bolivia, Ecuador, and the United Kingdom. We then analyzed the sequences of the PS-ID region, i.e., the intergenic region between the *rpl16* and *rpl14* genes present in the chloroplast genome, for all samples. A phylogenetic tree was then created by the maximum likelihood method using the above sequences (Fig. 7.3). Analysis of 525 bp fragment showed that the samples resolved into three clusters based on SNP differences at the 29th, 327th, and 328th bases. The A1 cluster includes samples with an A at position 29, T at position 327, and C at position 328, and this cluster included 61 of the 99 samples analyzed. Moreover, the A1 cluster contained wild taros (WT, marked with a red star) collected around Kutsukake hot spring in Nagano Prefecture, wild taros (WT, also marked with a red star) collected at Ryugenji Temple, Yamanashi Prefecture, and the "Eguimo," "Dodare," "Hasuimo," "Akame," and "Celebes" domesticated cultivars. These samples originated from cultivar groups such as the "Taira-ooba" and "Kurojiku" clades. These samples were collected from sites in Japan, China, Taiwan, South Korea, India, Nepal, Pakistan, Malaysia, the Philippines, American Samoa, Bolivia, Brazil, Costa Rica, and the United Kingdom. In addition, samples of unknown variety collected in Taiwan (CV:cultivar), India (CV), Hawaii (CV), and Tonga (CV) were also included. Importantly, all samples in the A1 cluster were triploid, and morphologically we observed a high proportion of var. *antiquorum*-type cultivars that produced many daughter and granddaughter corms.

Previous studies included six loci on the chloroplast genome (i.e., *rps16* intron-5′-*rps16* CDS-IGS toward *trnQ*, *trnY*-IGS-*trnE*, IGS between *trnT* and *psbD*, *rbcL*, *petD*-IGS-*rpoA*, IGS between *rpl22*, and *rps19*-*rps19*-IGS-*rpl2*) (Ahmed et al. 2020). Phylogenetic results based on these divided taros into three clades (i.e., CI, CII, and CIII; Fig. 7.3). The CI clade, which includes samples from Japan, India, Nepal, Bangladesh, Myanmar, Sri Lanka, the Philippines, Vietnam, Thailand, Australia, Easter Island, New Zealand, New Guinea, East Timor, Hawaii, the Société

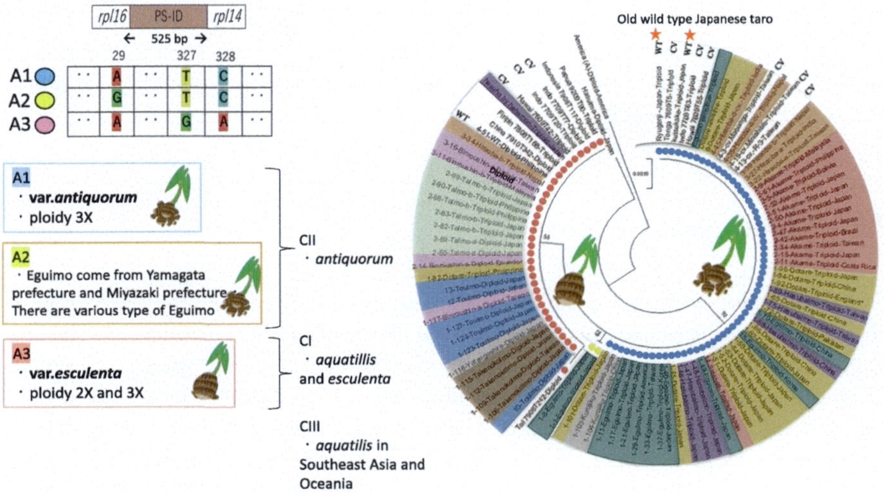

Fig. 7.3 Phylogenetic analysis based on sequence data for the PS-ID region of the chloroplast genome. Also shown is a comparison with the result of dividing the chloroplast genome into groups CI, CII, and CIII by six gene regions by Ahmed et al. (2020)

Islands, Fiji, Cyprus, Madagascar, and Egypt, includes wild and cultivated taro species collected from a wide range of sites from Oceania to Africa. Taros from this group were diploid and triploid, with small parental corms and long, vigorous stolons of the var. *aquatilis* and var. *esculenta* types. CII varieties belonged to the group of triploid cultivated taros collected from subtropical and temperate zones in Japan, India, Nepal, Pakistan, Sri Lanka, New Zealand, Madagascar, and Ethiopia. They showed morphological characteristics of the var. *antiquorum* type. In contrast, CIII taros were of the var. *aquatilis* type and were collected from tropical and subtropical regions including the Philippines, Myanmar, Thailand, Vietnam, Australia, and New Guinea. Moreover, we observed that *C. formosana* samples collected from China and Taiwan formed a distinct subclade. A similar clustering of partitioning into clades CI-CIII has been reported in restriction fragment length polymorphism (RFLP) analysis of the taro chloroplast genome (Ochiai et al. 2000).

Further sequence analysis of the PS-ID region of the taro samples revealed that samples showing the 327th base as G and the 328th base as A belonged to CI, while sample which had T as the 327th base and C as the 328th base belonged to CII or CIII (Tabuchi, unpublished data). In the present results, CI corresponds to the A3 clade, while CII corresponds to both A1 and A2 clades. Taken together, these results suggest that patterns of genetically maternal taro lineages match patterns associated with corm morphology (Figs. 7.1 and 7.3).

7.4 Phylogenic Analysis Using SNP Markers of the GRAS-Di Amplicon in Taro Varieties

Next, we performed further phylogenetic analyses on data from 96 samples of the "Eguimo," "Hasubaimo," "Dotare," "Taira-ooba," "Akame," "Taimo," "Tounoimo," "Takenokoimo," "Binroushin," "Kurojiku," and "Hasuimo" taro varieties, as well as from the *C. gigantia* and American taro (*Xanthosoma*) outgroups. Moreover, we also included unknown samples collected from sites in 21 countries, including Japan (33 samples), China (7 samples), Taiwan (10 samples), Korea (1 sample), India (10 samples), Nepal (2 samples), Pakistan (1 sample), the Philippines (6 samples), Malaysia (1 sample), Indonesia (6 samples), Thailand (3 samples), Brunei (1 sample), Tonga (2 samples), Hawaii (2 samples), Papua New Guinea (1 sample), Solomon Islands (1 sample), American Samoa (1 sample), Costa Rica (1 sample), Brazil (1 sample), Ecuador (1 sample) and the United Kingdom (1 sample). These samples were used as source material for GRAS-Di (Genotyping by Random Amplicon Sequencing-Direct) analysis.

GRAS-Di amplicon sequences were mapped to the reference genome (GenBank: GCA_014218235.1, Accession: PRJNA587719) using Burrows-Wheeler Aligner (BWA). SNPs were then identified using the Population feature implemented in Gstacks (parameters: min-mapq 25). We only used SNPs that were genotyped in >90% samples for subsequent analysis. Since our samples contained both diploid and triploid samples, pseudo-haploidization (parameters: pileupcaller, min-mapq 25, min-basequality 30) was performed to randomly select one allele from each sample. Genetically linked sites were then removed using PLINK version 1.9 (parameters: plink–indep 50 5 1.5), which ultimately yielded a total of 4872 SNPs (Purcell et al. 2007).

These 4872 SNPs were then used to create a phylogenetic tree using the NJ method. Subsequently, a PCA-UMAP analysis was performed with the number of principal components set to 4 and the number of nearest neighbor components to 50 (Fig. 7.4). These results identified an "Eguimo" cluster, a "Hasubaimo," "Dotare," and "Kurojiku" cluster, a "Akame," "Tounoimo," "Takenokoimo," "Binroushin," and "Taimo" clusters, a "Taira-ooba," "Dotare" cluster, and an unknown cluster. According to this analysis, varieties within each cluster had similar genotypes. Further analysis of the results of SSR marker analysis (Motohashi et al., unpublished) and the phylogenetic analysis of 2055 GRAS-Di amplicons showed that the "Dotare" variety was present in multiple clusters. Many "Dotare" varieties that were similar to "Hasubaimo" originated from Japan, while "Dotare" varieties that were more similar to "Taira-ooba" were collected from the UK or other sites outside Japan, including sites in China, Pakistan, and the Philippines. The life histories of the "Dotare" varied considerably, including early, medium, and late maturing plants, and seedling shape also varied from round to long oval. Overall, this group was highly diverse in morphology. We also identified many homologous and varieties present in Southeast Asia and East Asia, especially Taiwan. We speculate that "Dotare" may be scattered among multiple clusters because this group contains many similar varieties who were principally identified based on morphological

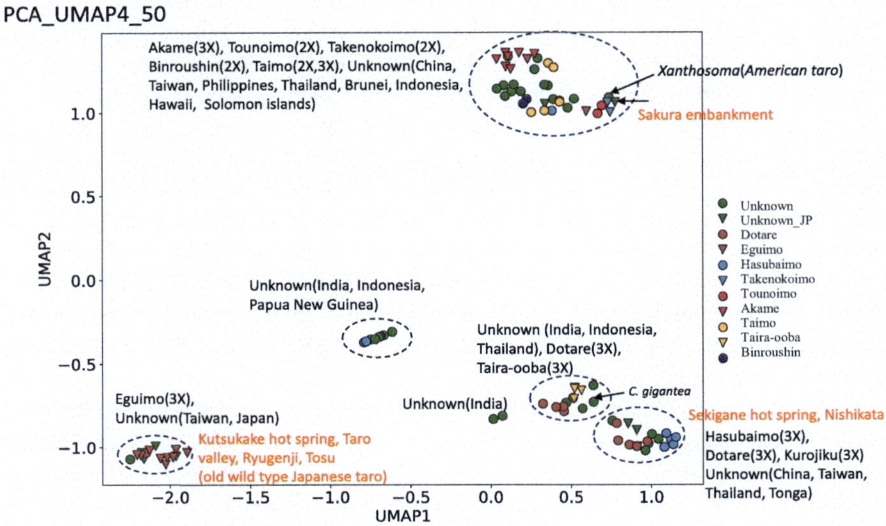

Fig. 7.4 Phylogenetic analysis using 4872 SNPs

characteristics. However, it has also been suggested that "Dotare" varieties grown in Japan may be genetically different from those found overseas.

The results of PCA-UMAP and phylogenetic tree analyses indicated that "Eguimo" occupied a genetically intermediate position between the "Akame" or "Tounoimo" group and the "Dotare" or "Hasabaimo" group (Fig. 7.4). In contrast, using phylogenetic trees based on the dominant markers of the 2055 amplicons analyzed here, the "Hasubaimo" and "Dotare" clusters were adjacent to the "Akame" cluster, and therefore the results of these two analyses were inconsistent. Accordingly, we performed f_4-statistical analysis using the 4872 SNPs to clarify the relationship between the "Eguimo," "Dotare," and "Akame" groups (Fig. 7.5). These results showed that "Eguimo" was most similar to "Hasubaimo"/"Dotare," while "Hasubaimo"/"Dotare" was slightly more similar to "Akame" than "Eguimo." Taken together, these results suggest that "Hasubaimo"/"Dotare" may have characteristics of "Akame" and "Eguimo." However, it can also be inferred that the "Hasubaimo"/"Dotare" varieties arose many centuries ago (or even earlier), before diploid and triploid varieties had mixed and the differentiation of cultivar groups had progressed. They would have emerged via crossing diploid varieties that were genetically close to "Eguimo" with diploid varieties that were genetically close to "Akame." Subsequently, triploids, which are high-yielding, increased in both variety groups, thereby forming the present cultivar groups.

7.5 The Origin of Taro

SNP analysis of GRAS-Di amplicons revealed that wild-type samples of unknown variety and cultivated samples collected in India did not belong to known variety groups and formed their own unique clusters. The tendency of these samples to form clusters earlier than other varieties suggested that these samples were among the oldest lineages. Thus, we infer that taro originated in Northeastern India. In addition, we also conclude that the "Binlousin," "Takenokoimo," "Tounoimo," and "Yatsugashira" cultivars, which have been cultivated in China and Taiwan as important varieties for a long time, had different developmental routes from a common Indian origin. "Eguimo" is known as an old variety group, but the native taros used in this study, which were collected around Kutsukake hot spring, Nagano Prefecture (Fig. 7.2a), Ryugenji Temple, Yamanashi Prefecture (Fig. 7.2b), Tosu City, Saga Prefecture, and Taro Valley in Izumo City, Shimane Prefecture (Fig. 7.2c), were all identified as "Eguimo," suggesting that the "Eguimo" variety group was introduced into Japan relatively early. Similar strains were found to have spread to Taiwan, Korea, China, and throughout Oceania. Since taro originated in India and the Japanese archipelago was connected by land to the Eurasian continent at that time, it may be possible that taro was introduced from India by Yaponesians via Taiwan and Korea (Fig. 7.6). In Korea, trandang (taro soup) is a characteristic autumn dish, and the Amis of Taiwan traditionally cook taro, millet, and rice together, which resembles the "chimaki" dish of Kyoto.

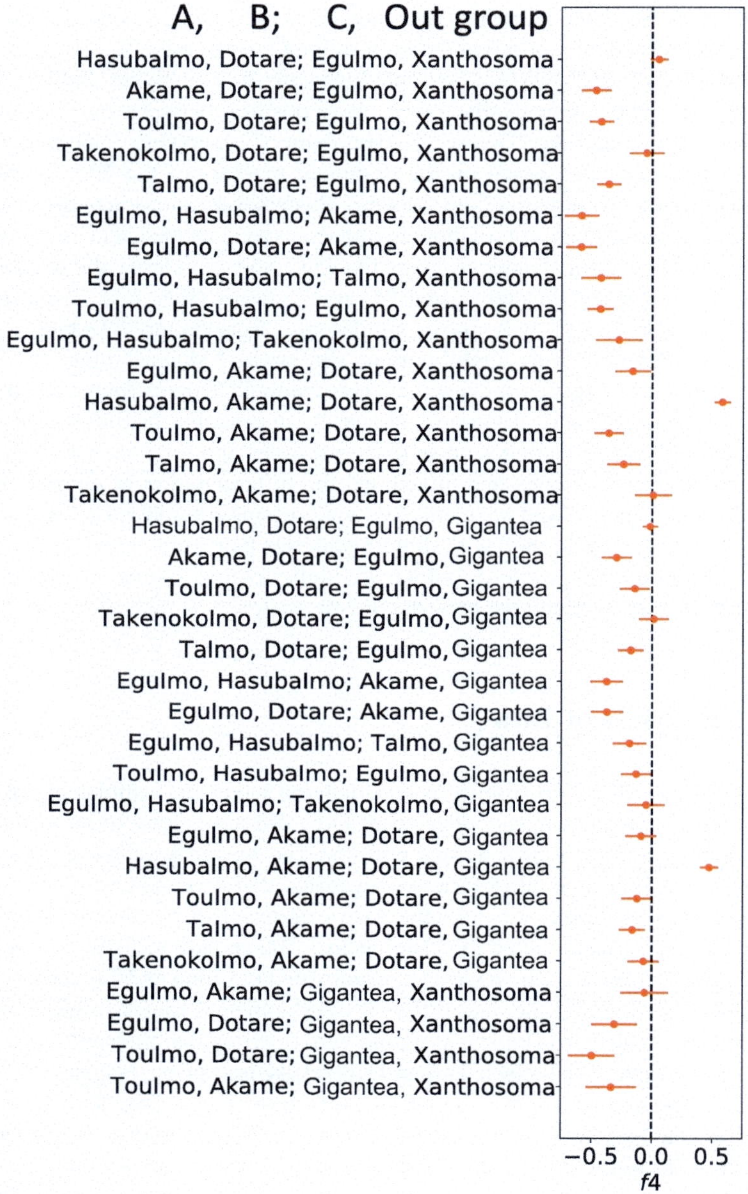

Fig. 7.5 f_4 analysis using 4872 SNPs. If A and C are close, red spot shows more positive than 0. If B and C are close, it shows more negative than 0

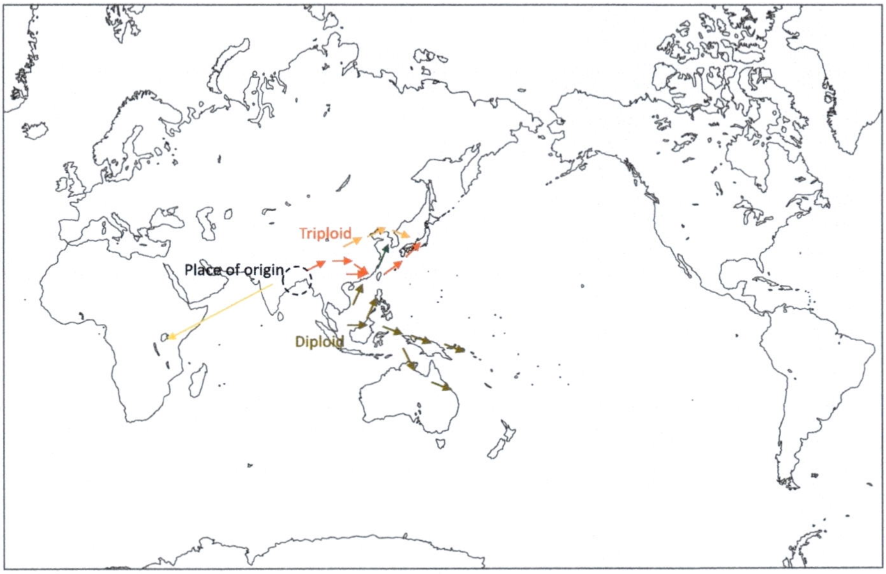

Fig. 7.6 Predicted origin region and propagation pathways of taro

"Eguimo," which has been growing in the Japanese wilderness for many years, and the "Dotare" varieties that are widely cultivated in Japan are both triploid. Interestingly, taro varieties cultivated in Southeast Asia and Polynesia are mostly diploid. It is thought that the diploid and triploid varieties separated relatively early from the Northeastern Indian sites of taro origination, after which they were spread by humans. Moreover, a small number of triploid taro plants have been produced by crossbreeding or self-pollination of wild diploid taro plants in Bangladesh; thus it is hypothesized that triploid taro may have originated from diploid hybrids or by self-pollinated cultivars (Isshiki et al. 1999).

Competing Interests The authors have no relevant financial or non-financial interests to disclose.

References

Ahmed I, Lockhart PJ, Agoo EMG, Naing KW, Nguyen DV, Medhi DK, Matthews PJ (2020) Evolutionary origins of taro (*Colocasia esculenta*) in Southeast Asia. Ecol Evol 10(23):13530–13543

Bellinger MR, Paudel R, Starnes S, Kambic L, Kantar MB, Wolfgruber T, Lamour K, Geib S, Sim S, Miyasaka SC, Helmkampf M, Shintaku M (2020) Taro genome assembly and linkage map reveal QTLs for resistance to taro leaf blight. G3 (Bethesda) 10(8):2763–2775

Helmkampf M, Wolfgruber KT, Bellinger MR, Paudel R, Kantar BM, Miyasaka CS, Kimball LH, Brown A, Veillet A, Read A, Shintaku M (2018) Phylogenetic relationships, breeding implica-

tions, and cultivation history of Hawaiian taro (*Colocasia Esculenta*) through genome-wide SNP genotyping. J Hered 109(3):272–282
Isshiki S, Otsuka K, Tashiro Y, Miyazaki S (1999) A probable origin of triploids in taro (*Colocasia esculenta* (L.) Schott). J Jpn Soc Hortic Sci 68(4):774–779
Kumazawa S, Niuchi K, Honda F (1956) Classification of taro varieties in Japan. Jpn Soc Hortic Sci 25:1–10. (In Japanese with English summary)
Matthews PJ (1991) A possible tropical wildtype taro: *Colocasia esculenta var. aquatilis*. Indo-Pacific Prehistory Assoc Bull 11:69–81
Matthews PJ, Ghanem ME (2021) Perception gaps that may explain the status of taro (*Colocasia esculenta*) as an "orphan crop". Plants People Planet 3(2):99–112
Ochiai T, Tahara M, Yoshino H (2000) Phylogenetic relationships of taro and allied species based on restriction fragment length polymorphisms (RFLPs) of chloroplast DNA. Sci Rep Fac Agric Okayama Univ 89(1):15–21. (In Japanese with English summary)
Plucknett DL (1983) Taxonomy of the genus Colocasia. In: Wang J-K (ed) Taro: a review of Colocasia esculenta and its potentials. University of Hawaii Press, Honolulu, pp 14–33
Purcell S, Neale B, Todd-Brown K, Thomas L, Ferreira MA, Bender D, Maller J, Sklar P, de Bakker PI, Daly MJ, Sham PC (2007) PLINK: a tool set for whole-genome association and population-based linkage analyses. Am J Hum Genet 81(3):559–575
Rao VR, Hunter D, Eyzaguirre PB, Matthews PJ (2010) Ethnobotany and global diversity of taro. In: Rao VR, Matthews PJ, Eyzaguirre PB, Hunter D (eds) The global diversity of taro: ethnobotany and conservation. Bioversity International, Rome, pp 1–5
Takeshita A, Miyazaki S, Tashiro Y, Yanagawa M, Tabaru M (1991) A clone of taro (*Colocasia esculenta* Schott) growing naturally in Tosu City of Saga prefecture in Japan. Bull Fac Agric, Saga Univ 71:113–122. (In Japanese with English summary)
Yin J, Jiang L, Wang L, Han X, Guo W, Li C, Zhou Y, Denton M, Zhang P (2021) A high-quality genome of taro (*Colocasia esculenta* (L.) Schott), one of the world's oldest crops. Mol Ecol Resour 21(1):68–77

Chapter 8
History of Propagation and Genetic Structures in Three Japanese Millets: Proso Millet, Foxtail Millet, and Barnyard Millet

Kazuhiro Satomura

Abstract Currently, rice is a staple food in Eastern Eurasia, but historically, millet has also been an important grain. Proso millet, foxtail millet, and Japanese barnyard millet were domesticated from wild plants in East Asia. Cultivation of these millets expanded over large areas of the world because they can adapt to colder regions at higher latitudes than other cereals. Millet cultivation has historically been as important as rice cultivation in Yaponesia because the northern limit of rice cultivation is located within the country. In northern Yaponesia, Japanese barnyard millet might have been in use by 7000 BC and domesticated by 2700 BC. At approximately 1000 BC, rice, proso millet, and foxtail millet were introduced into Yaponesia from the Eurasian continent. Subsequently, rice has become the staple food of the Yaponesian population, but the three aforementioned millets have also been traditionally important in Yaponesia, especially as the main carbohydrates in the diet of northern Yaponesia. These three millets have evolved in East Asia through domestication via unconscious selection regarding farming and food processing methods. Furthermore, evolution of the response to day length is essential for these millets to adapt to different latitudinal environments, which would have influenced their historical propagation.

Keywords Agricultural history · Domestication · Comparative genomics · Archeology · Local adaptation

K. Satomura (✉)
Department of Bioscience, Nagahama Institute of Bio-science and Technology, Nagahama, Japan
e-mail: k.satomura@nagahama-i-bio.ac.jp

N. Osada et al. (eds.), *Phylogeographic History of Plants and Animals Coexisting with Humans in Asia*, Evolutionary Studies,
https://doi.org/10.1007/978-981-97-6887-5_8

8.1 Introduction

Cereals have played a crucial role as a major source of carbohydrates in human societies since the dawn of agriculture, which began in the early Holocene in parallel spreading from tropical to temperate zones globally. Currently, maize, wheat, and rice dominate global cereal production; however, historically, various cereals have been cultivated by different cultures worldwide. The other cereals used by humans are called "millet." The cereals considered "other" vary according to culture. Although cereals are mainly crops in the *Poaceae* family (Fig. 8.1), they can sometimes include crops such as quinoa, which is an *Amaranthaceae* plant, and buckwheat, which is a *Polygonaceae* plant. Millet in English often refers to minor grains belonging to *Chloridoideae* or *Panicoideae* of the *Poaceae* family.

Millet has been cultivated in large areas of the world where rice cannot be grown in the summer because of its ability to adapt to a sterile environment. In photosynthetic processes, millets perform C_4 carbon fixation, in contrast to the C_3 carbon fixation typical in common plants. C_4 plants can grow in semi-arid, nutrient-poor, and high-temperature areas because they use water and nitrogen more efficiently. Therefore, it remains an important cereal in the semi-arid regions of South Asia and around the Sahara. Millet contains approximately the same amount of energy,

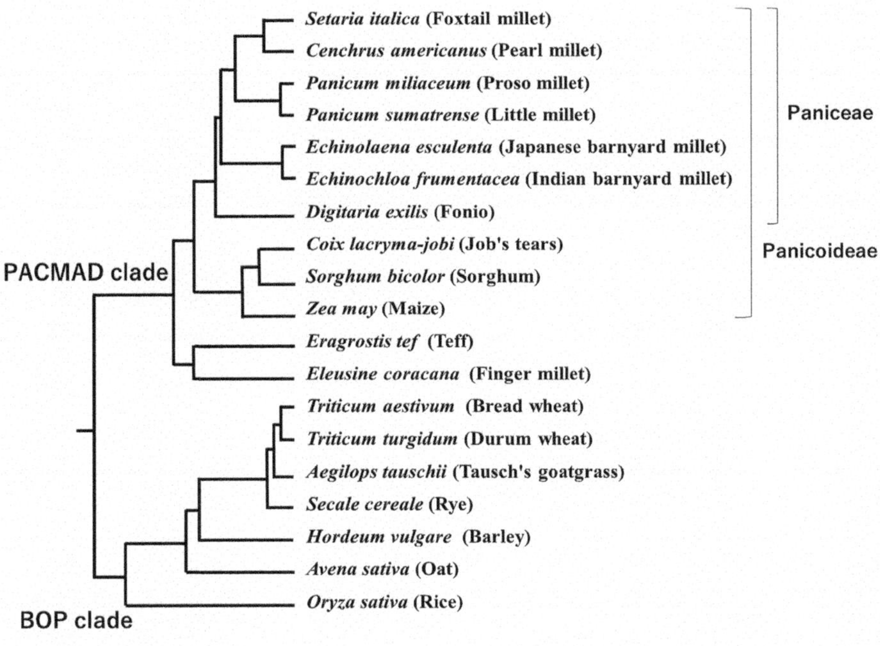

Fig. 8.1 Phylogenetic relationships of the primary edible Poaceae crops. Poaceae plants are broadly classified taxonomically into the PACMAD clade and BOP clade

protein, and minerals as wheat, is high in fiber, and is gluten free. It is also used worldwide not only for human consumption, but also as livestock feed and ornaments.

Rice is now a staple food in East Asia, Southeast Asia, and South Asia, although it was originally a tropical crop that was not adapted to cold climates. In Japan, rice was first cultivated in Hokkaido in 1692, but its cultivation did not become widespread until 1873, when Kyuzo Nakayama produced the improved variety Akage. Historically, in northeastern Asia, foxtail millet (*Setaria italica*), proso millet (*Panicum miliaceum*), and barnyard millet (*Echinochloa esculenta* and *E. frumentacea*) were grown in summer, whereas wheat and barley were cultivated in winter. In China and Japan, the main grains are called gokoku (meaning five grains); the grains that are included in gokoku vary according to time and region, but most of them are selected from rice, wheat or barley, proso millet, foxtail millet, barnyard millet, and legumes. Millets, along with rice, are still used currently in the traditional ceremony of Japan known as the Niinamesai, also called Shinjosai, in which the Emperor celebrates the harvest of the gokoku; this day is now a national holiday. This chapter reviews the history of the three major Yaponesian millets, proso millet, foxtail millet, and barnyard millet.

8.2 History of Proso and Foxtail Millets

From Wild to Domesticated Proso Millet

Proso millet is a plant of genus *Panicum*. The genus *Panicum* comprises around 450 species, which are distributed in the tropics and subtropics, and grow wild in grasslands, farmlands, wastelands, and beaches. The larger species have branched inflorescences that can reach over 60 cm, and their ears sparsely bear large numbers of small seeds. The basic chromosome number of the genus *Panicum* is $x = 9$, however there are many species that have more than one pair of homologous chromosomes set; i.e., polyploidy. A polyploid genome is composed of redundant chromosome sets (subgenomes) that can originate from different species and even arise within a single species. Most angiosperms have experienced one or more polyploidization events during their evolution (Masterson 1994). And it can also occur during the molecular evolutionary process of plant domestication. Generally, plants of the genus *Panicum* range from diploid to dodecaploid. For example, switchgrass (*P. virgatum*) carries a tetraploid or octoploid genome with each individual. The switchgrass subgenomes diverged from a common ancestor 6.7 million years ago and hybridized again 4.6 million years ago (Triplett et al. 2012). This perennial grass has been bred as a forage crop, but its high biomass yield and ability to adapt to various environments have rendered it a model for bioenergy feedstock crop since the 1990s. Its duplicated genomes are thought to provide genetic diversity to switchgrass and helped its adaptation to its environment.

Three species in the genus *Panicum* are known to have been historically domesticated for food worldwide: *P. miliaceum*, *P. sumatrense*, and *P. sonorum*. *P. miliaceum*, also known as broomcorn, proso, or common millet, is the most widespread domesticated cereal in the genus *Panicum* worldwide (Fig. 8.2a). *P. miliaceum* has only an allotetraploid (2n = 4x = 36) genome with a size of approximately 900 Mb, and its subgenomes diverged about 5.9 million years ago and later tetraploidized (Hamoud et al. 1994; Hunt et al. 2014; Shi et al. 2019). The origin of the two subgenomes of *P. miliaceum* was investigated using sequence homology, which suggested that one of them is related to the diploid genome of witchgrass (*P. capillare*) and the other is related to one of the subgenomes of torpedo grass (*P. repens*) (Hunt et al. 2014) (Fig. 8.2b). *P. capillare* is a small grass that grows on roadsides and farmland and is native to North America, which is a colder habitat compared to the other *Panicum* species. It has recently expanded its distribution area throughout Eurasia. There is a record of the excavation of its seeds from a site in Caucasus of 5000 BC (Lisitsyna and Prishchepenko 1977), however this identification is unlikely to be correct, as it is believed that *P. capillare* was not distributed in that location at that time. Although *P. repens* is found throughout the world in tropical and subtropical coastal areas, it is considered as a naturalized plant in most regions and its exact origin is not known. *P. repens* encompasses a variety of diploid to decaploid individuals based on both x = 9 and x = 10. However it is not certain whether the polyploid individuals are fertile, as only propagation by rhizome has been confirmed (Majure 2009). Both *P. capillare* and *P. repens* are now naturalized in Yaponesia, although the former is distributed in colder regions, such as Hokkaido, whereas the latter is found in warmer regions, such as Okinawa; however, the distribution areas of the two species do not overlap. Because it is difficult to reconcile the reproductive mode, distribution range, and ploidy of these two species, it is unlikely that they are direct ancestors, although they are related to the ancestor of *P. miliaceum*. *P. sumatrense*, commonly known as little or sama millet, is cultivated in South and Southeast Asia and is believed to have been domesticated from a wild species during the Indus civilization around 3000 BC (de Wet et al. 1983). Moreover, large quantities of seeds have been excavated from the Harappa site (Weber and Kashyap 2016). It typically has an allotetraploid (2n = 4x = 36) genome, and one subgenome is related to *P. repens*, similar to *P. miliaceum* (Hunt et al. 2014). Wild *P. psilopodium* can

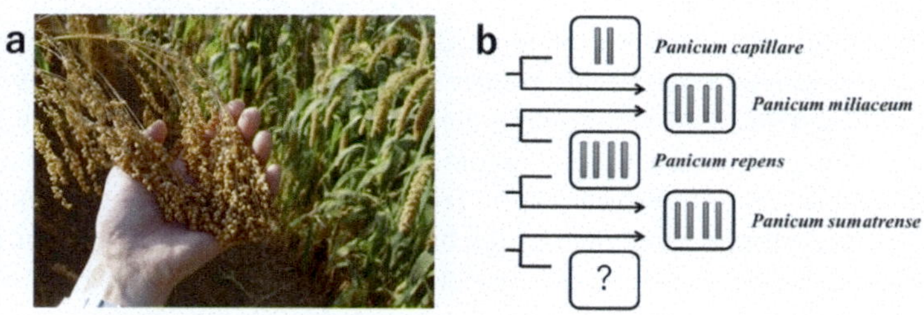

Fig. 8.2 (**a**) *Panicum miliaceum* (courtesy of Shin'ichiro Fujio). (**b**) Relationships of the genomes of *P. miliaceum* and *Panicum sumatrense* with the genomes of wild *Panicum* species

hybridize with *P. sumatrense* therefore this plant is the ancestral wild species, related to the ancestor of *P. sumatrense*, or a feralized *P. sumatrense* (Hiremath et al. 1990). *P. sonorum* is a native crop that has been cultivated and domesticated in the Sonoran Desert, which stretches across the border between the United States and Mexico; however, it has not been studied in detail biologically nor archeologically (Nabhan and de Wet 1984).

The ancestral wild species and evolution of *Panicum miliaceum* are not fully elucidated. However, past surveys have reported two weedy proso millet subspecies. It was first described in Manchuria in northeastern China as *P. millaceum* subsp. *ruderale* because it has the same characteristics as cultivated proso millet, with the exception of an erect ear, a narrow lemma, and seed shattering (Kitagawa 1937; Scholz 1983). Since then, weedy proso millet has been reported in large areas of Eastern Europe, the Aralo–Caspian Basin, Buxoro in Central Asia, Siberia, and North America. The weedy proso millet found in Europe and North America exhibits characteristics that are intermediate between the cultivated *P. miliaceum* and the weedy *P. miliaceum* subsp. *ruderale* and was named *P. miliaceum* subsp. *agricolum*. It is considered feral from cultivated *P. miliaceum*, and its herbicide resistance renders it a threat in modern agriculture. A recent population genome analysis revealed that this weedy subspecies is related to Chinese landraces of *P. miliaceum* and is more ancestral, as assessed based on molecular phylogenetics (Chen et al. 2023). The observations of gene flow between cultivated and wild populations suggested that *P. miliaceum* subsp. *ruderale* is an ancestral wild species with postdomestication gene flow.

The region of the original *P. miliaceum* domestication is difficult to determine based on the distribution range of the ancestral wild species, because this datum remains unclear. However, molecular phylogenetic inference indicated that the genome sequences of Chinese landraces were most closely related to wild subspecies and ancestral types (Chen et al. 2023; Li et al. 2021). Because domesticated plants should exhibit the highest diversity at their place of origin, the diversity center is considered to correspond to the place of origin. Classical studies performed by Vavilov and Harlan have also identified the origin of *P. miliaceum* in East Asia, which is its largest diversity center. However, several researchers have proposed multiple origins for *P. miliaceum*, as it sometimes appears to have diversity centers in Europe and South Asia. Nevertheless, population genome analyses indicate that the diversity center of this species is located in northern China, and that the European lineage is derivative (Chen et al. 2023; Li et al. 2021). The oldest archeological evidence of *P. miliaceum* in the world comes from grains found in excavated storage pits at the Neolithic Cishan site located on the northern Yellow River, which have been dated using radiocarbon ^{14}C to 8400–6800 BC (Lu et al. 2009). Large quantities of carbonized grains before 5500 BC have also been excavated from neighboring sites, such as the Dadiwan site on the Loess Plateau and the Xinglonggou site in eastern Inner Mongolia (Barton et al. 2009). These areas are located inland, have a low rainfall, strong winds, and cool temperatures; and lack water and soil nutrients (Li et al. 2009). *P. miliaceum* stands out among cereals for its lowest water and nutrient demands, and its short growing time (maturation occurring within 60–100 days after sowing) contributes to its high value in the region. *P. miliaceum* has been found as carbonized grains and grain impressions in pottery at many sites in Eastern Europe and the Mediterranean and Black Sea coastal regions before 5000 BC. However, excavated small grains are not always well preserved, which hampers the identification of plant species. The reconsideration of the dating of grains excavated from the Caucasus sites suggests estimates dating back to 2000 BC (Martin et al. 2021). Moreover, recent genomic and archeological evidence rarely supports a non-Chinese origin site. Overall, it is reasonable to assume that north-eastern China, where the Cishan site is located, was the original site of *P. miliaceum* domestication.

Domestication is a mutualistic relationship between humans and crops that extends beyond the temporary cultivation of wild grass. Humans receive resources and/or services in exchange for maintaining the agricultural environment and managing their reproductive succession from one generation to the next (Zeder 2015). As a result, co-evolution occurs, and crops acquire traits that are adapted to the human agro-environment and needs. *P. miliaceum* grains excavated from sites located in northern China were approximately 1 mm in size at 6000 BC, and gradually increased in length, width, and thickness by approximately 1.5 times over the subsequent 3000 years (Stevens et al. 2021). Furthermore, a comparative genomic analysis indicated that *P. miliaceum* evolved during the domestication process via the loss of seed shattering, changes in inflorescence and seed color related to plant adaptation to the environment and nutrient accumulation, and changes in ear shape to increase grain yield (Chen et al. 2023). These changes are common with the trait changes that occur during the domestication of grains and legumes in general (Abbo et al. 2014; Fuller 2007). In East and Southeast Asia, porridge, steaming and mochi, i.e., rice cake, are methods used for processing grains, and East-Asian landraces of *P. miliaceum*, particularly in Yaponesia, carry mutations in the *Waxy* gene, leading to the production of viscous starch (Araki et al. 2012). Domesticated *P. miliaceum* has 37% less genetic diversity compared with wild species (Chen et al. 2023). The genetic diversity of the wild species is only one-fifth that of wild rice and soybean, suggesting that either the ancestral population was distributed in a small area, or the wild species was greatly affected by human activities. Transposon insertions can alter gene expression levels, which may have been beneficial for *P. miliaceum* domestication (Chen et al. 2023). In contrast, subgenomes with more transposons lost more genes. The redundancy in the genes from two subgenomes would have contributed to one-gene degeneration during the evolution of *P. miliaceum*.

From Wild to Domesticated Foxtail Millet

Foxtail millet is a plant of genus *Setaria*. Most species of the genus *Setaria*, with over 100 recorded species, including foxtail millet, are distributed from the tropics to the subarctic, growing wild in grasslands, farmlands, and roadsides. *Setaria* species are generally annual grasses, produce dense cylindrical or branched inflorescences, and have larger ears that reach more than 15 cm in length and are covered with the bristles of small ears. Although the seeds of wild *Setaria* species are smaller than those of domesticated species, they produce hundreds of dense seeds from a single inflorescence and are self-compatible. Various species of wild *Setaria* plants are edible in Asia, Africa, and Pacific island countries. *Setaria* plants have a basic chromosome number of x = 9, like *Panicum* plants, with many ploid species. Although *S. pumila* (yellow foxtail) and *S. viridis* (green foxtail) are native to Eurasia, they have spread throughout the world from the temperate zone to the tropics and subarctic. *S. glauca*, which is morphologically very similar to *S. pumila*, is also called yellow foxtail and is believed to be distributed worldwide, however, species identification may be confusing. At least those found in Yaponesia are *S. pumila*. *S. parviflora* (marsh bristlegrass), which is native to the warmer regions of North America, can grow in dry, iron-rich soils and has extended its distribution throughout Asia and the islands of the Pacific Ocean and Indian Ocean. *S. verticillata* (bristly foxtail) and *S. palmifolia* (palmgrass) are distributed in tropical and warm temperate areas and are widespread worldwide. *S. faberi* (giant foxtail) was originally distributed widely in the East-Asian area including Yaponesia, although its distribution has recently expanded to North America. *S. faberi*, *S. pumila*, and *S. viridis* encompass various agricultural herbicide-resistant strains within their species, with strains adapted to modern human agricultural environments sometimes causing serious agricultural damage as weeds (Stoltenberg and Wiederholt 1995; Wang and Dekker 1995).

S. italica, commonly known as foxtail millet, is cultivated worldwide in areas that are almost the same as *P. miliaceum*. *S. italica* has only one diploid (2n = 2x = 18) genome with a size of approximately 450 Mb. This is the second smallest genome after rice among the major domesticated cereals in the Poaceae, which facilitate whole-genome analysis. The ancestral wild species of *S. italica* was identified as *S. viridis*, a diploid annual grass, based on morphological and cytological evidence, as well as RAPD, AFLP, and other genetic markers (Fig. 8.3) (Li et al. 1945; Kihara and Kishimoto 1942). *S. faberi* is a larger-sized, closely related species that overlaps with the distribution area of *S. viridis* but has a tetraploid genome. *S. viridis* is a self-compatible annual grass with a small plant size and various environmental adaptations, including drought resistance; moreover, the size of its wild seeds can exceed 2 mm. Therefore, it has been proposed as a model for C_4 plants that have potential as bioresources, such as biofuels (Mamidi et al. 2020). Environmental adaptation in *S. viridis* is attributed to its high phenotypic plasticity, although its genetic diversity is not very high (Doust and Kellogg 2006; Wang et al. 2010). Small genetic diversity can be interpreted to mean that its effective population size is small. The distribution of *S. viridis* in Eurasia is thought to have been much more restricted compared to what it is currently.

The place of origin of *S. italica* domestication is also difficult to predict based on the past distribution area of *S. viridis*. Because *S. italica* is frequently excavated from the same sites as *P. miliaceum*, its place of origin has been debated alongside *P. miliaceum*'s. *S. italica* was thought to have the same East-Asian origin as *P. miliaceum*, its largest diversity center; however, other researchers have suggested multiple origins in Europe and South Asia. Furthermore, molecular phylogenetic

Fig. 8.3 (**a**) *Setaria italica* (courtesy of Shin'ichiro Fujio). (**b**) *Setaria viridis*. (**c**) *Setaria pycnocoma*. (**d**) Phylogenetic relationships among *Setaria* species. The black arrow indicates domestication

analyses of *S. italica* have identified small diversity centers in South Asia and from Eastern Europe to Central Asia, in addition to East Asia, which has the greatest diversity (Fukunaga et al. 2002b; Hirano et al. 2011). These lineages appear to have diverged almost simultaneously, and the Eastern-European and Central-Asian lineages are often closely related to *S. viridis*. However, domesticated plants often hybridize with wild species or with other domesticated strains during their history, in which case the most ancestral cluster on the molecular phylogenetic tree may not be truly ancestral. The oldest archeological evidence of *S. italica* in the world stems from grains found in excavated storage pits at the Cishan site, which included *P. miliaceum*, although it appeared after 6800 BC (Lu et al. 2009). This implies that *P. miliaceum* was cultivated earlier than *S. italica* when agriculture began in northeastern China in the Neolithic Age. Since then, it has been suggested that *S. italica*, together with *P. miliaceum*, has supported human and livestock diets in northeastern China for over 2000 years (Barton et al. 2009). Phenotypic changes in *S. italica* caused by domestication continued to occur in China for more than 2000 years (Stevens and Fuller 2017). Subsequently, *S. italica* and *P. miliaceum* were excavated from sites with a more recent date located throughout Eurasia. *S. viridis* is a companion weed of *S. italica*, with spontaneous interspecific hybridization occurring in both directions at rates ranging from 0.002% to 0.6% depending on plant density and distance (Till-Bottraud et al. 1992). This gene flow is a confounding factor in molecular phylogenetic analyses. Thus, the phylogenetic analyses suggesting that the Eastern European and Central Asian *S. italica* lineages sometimes appear to diverge slightly older than the Chinese lineages, for which there is the oldest archaeological evidence, may be the result of gene flow from *S. viridis*. Consequently, northeastern China, the same area indicated for *P. miliaceum*, is the most reliable place of origin of *S. italica*.

The first reference genome of *S. italica* (Yugu1 strain) was released in 2015 as early as in millet. For *S. italica*, research platforms, including various omics data and genetic variants, for breeding and basic research, were available from an early time (Bennetzen et al. 2012; Li et al. 2022; Mamidi et al. 2020; Zhang et al. 2012). The size of grains from *S. italica* increased before other traits changed during the first 500–1000 years of cultivation (Fuller 2007). The increase in grain size continued after the establishment of the agricultural economy in the Yangshao period after 5000 BC, suggesting that domestication proceeded over a long period (Stevens and Fuller 2017). A comparative genomic analysis suggested that genes for grain yield, grain size, plant height, tiller number, heading date, drought resistance, cold resistance, heat resistance, carotenoid content, sugar content, chlorophyll content, photosynthetic capacity, and millet color have been selected during the domestication process of *S. italica* (Li et al. 2022). The *Waxy* gene in *S. italica* was suggested to have been lost independently in Yaponesia-Taiwan, Korea, and Southeast Asia via parallel transposon insertion (Fukunaga et al. 2002a). *S. italica* has lost approximately half of its genetic diversity compared with *S. viridis* (θ_w = 0.0027 and θ_w = 0.0059, respectively), and they share 75% of genetic variation (Wang et al. 2010). The genetic diversity of *S. italica* is at the same level as that of rice and sorghum. Although *S. italica* lost genetic diversity because of bottlenecks and artificial selection occurring during domestication, its ability to interbreed with *S. viridis* would have been a factor in maintaining its genetic diversity.

History of Proso and Foxtail Millet Propagation in Eurasia

Before the Neolithic Age, humans lived as hunter–gatherers. In East Asia, rice cultivation began in the Yangtze River basin and proso millet cultivation began in the Yellow River basin after 8000 BC. Archeological evidence indicates that for the first 1000 years, millet cultivation relied exclusively on proso millet. The subsequent addition of foxtail millet provided a sufficient yield to feed people and livestock (Barton et al. 2009). Although there are early traces of the use of proso and/or foxtail millet throughout the Yellow River basin and in Xinglongwa in eastern Inner Mongolia, this remained a hunter–gatherer society, with the agricultural society not being established until about 5000 BC (Stevens and Fuller 2017). Until then, the grain size of millet was close to that of the wild species; acorns and other foraged foods were collected seasonally, and isotopic data revealed that millet represented less than half of the staple diet. In the Middle Yellow River region and the Inner Mongolia region, there were agrarian cultures that continuously cultivated millets using cultivation tools, whereas in other areas, millet cultivation could not have been continuous. Moreover, probably unconscious artificial selection for important domestication traits in millets continued until after 5000 BC. Millets first increased in grain size, and agronomically important traits, such as loss of seed shattering, loss of seed dormancy, and synchronization of tillering and ripening, were fixed over time via unconscious selection. Since the beginning of agriculture in China, millets have been cultivated at high latitudes, north of the Yellow River in cold and dry climates, whereas rice has been cultivated at low latitudes, south of the Yangtze River in warm and humid climates. In the area located between the Yellow River and the Yangtze River, both millets and rice have been cultivated. After 3500 BC, human population migrations or cultural exchanges led to the spread of millet cultivation areas. Excavations of farming tools suggest that millet cultivation was widespread in the eastern Jilin and Heilongjiang provinces of northern China between approximately 4000 BC and 3000 BC. The triangulation of linguistic, archeological, and genetic evidence explains the spread of millet cultivation from the Yellow River basin to the Korean Peninsula and the Amur River basin in Primorsky Krai after 3500 BC because of migrations of agricultural populations with complex demographic dynamics (Robbeets et al. 2021). Millets, along with language and ritual culture, spread from Shandong Province along the sea route to Taiwan around 3000 BC, accompanied by rice (Sagart et al. 2018). Although all three grains are still cultivated in traditional Taiwanese practices, *S. italica* is culturally believed to be sacred by most of them. Excavated artifacts, such as pottery, suggest existence of cultural contact between Taiwan and the northern Philippines between 2000 and 1500 BC (Hung 2005). A genetic analysis of *S. italica* landraces indicated that strains from Taiwan and Philippines belong to the same group (Hirano et al. 2011). Therefore, it is possible that millets from Philippines were introduced via Taiwan.

Around 2000 BC, wheat and barley, which were domesticated in the Levant, spread to China (Dong et al. 2017). In Xinjiang, there are archaeological records of proso millet, foxtail millet, wheat, and barley cultivation from 3000 to 1300 BC (Li 2021). Later, the millets reached Central Asia by 1500 BC, and by 500 BC, they had spread throughout Eurasia (Fig. 8.4) (Stevens et al. 2016). *P. miliaceum* and *S. italica* were cultivated in almost the same areas until about 2500 BC. In turn, after 1500 BC, *P. miliaceum* was frequently excavated from sites located at higher latitudes and in colder areas, including Siberia and Europe. *P. miliaceum* would have been better adapted to the colder, drier regions of the high latitudes compared with *S. italica*. *S. italica* was frequently excavated from sites located at lower latitudes and in warmer areas, including Southeast Asia and South Asia. In those areas, *P. sumatrense* has been cultivated. *P. miliaceum* may not always have been necessary in areas where *P. sumatrense* is common. Most of the genetic diversity of *P. miliaceum* is found in East Asia, and only a few strains have propagated westward (Chen et al. 2023). This pattern appears to be common in *S. italica* (Hirano et al. 2011). The worldwide distribution of genotypes of *S. italica* landraces appear to be largely consistent with dispersal patterns determined using archeological dating, with the group of Southeast Asia, South Asia, and Africa and the group of West Asia to Europe, in addition to the East-Asian diversity center. A SLAF-Seq analysis of cultivated and wild proso millets suggested that the westward dispersal route from China included a high-latitude route via Mongolia to Europe and a midlatitude route via Xinjiang to South Asia and the Levant (Li et al. 2021). This is a compelling hypothesis considering the archeological dating record and the adaptation of millet to different environments at different latitudes. However, even if a large number of different millet strains arrived randomly in these regions, it is possible that strains

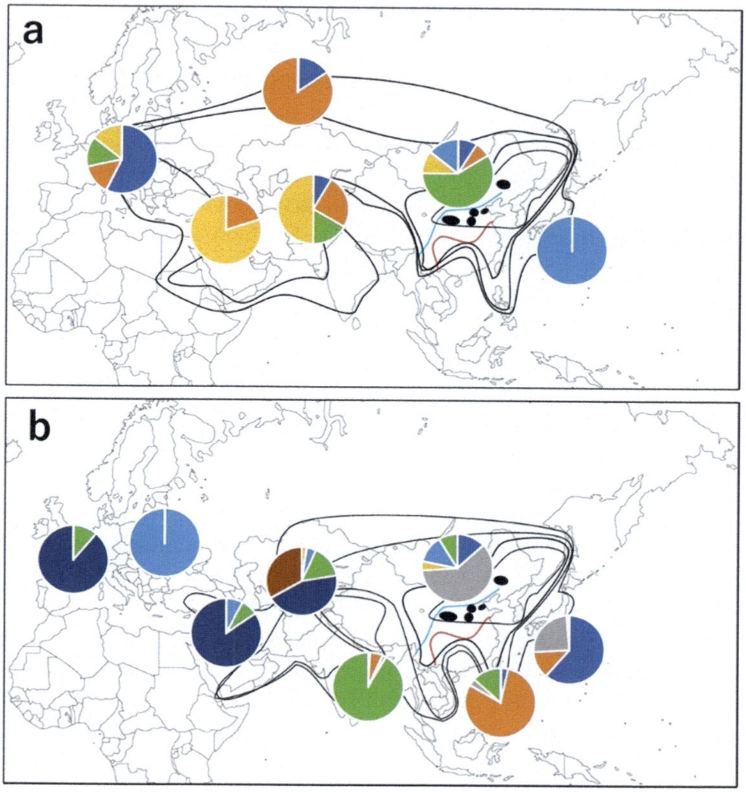

Fig. 8.4 For (**a**) *Panicum miliaceum* and (**b**) *Setaria italica*, the initial extent of cultivation before 5000 BC is indicated in black. The northern edge of rice cultivation (blue line) and the southern edge of millets cultivation (red line) before the spread of millet are presented (He et al. 2017). The contour lines denote the area of spread after 3500 BC by 1000 years. Modified from Stevens et al. (2016). In addition, pie charts present the percentages of landrace genotypes for *P. miliaceum* (Li et al. 2021) and *S. italica* (Hirano et al. 2011)

with similar genetic backgrounds survived because of environmental selective pressures; therefore, more multifaceted and detailed analyses are required to address this issue.

P. miliaceum subsp. *ruderale*, which is a weed proso millet that is a possible ancestral wild species of *P. miliaceum*, and *S. viridis*, which is the ancestral wild species of *S. italica*, have low genetic diversity relative to their wide distribution areas. The ancestral wild species of these two millets would have been distributed in a more restricted area, but their range expanded with the spread of cultivated species and human activities. In recent years, *S. viridis* has invaded North America and its distribution area is expanding rapidly. *S. viridis* in North America was derived from Europe, and its population structure was shaped depending on the latitude and climate within North America (Mamidi et al. 2020). Latitude is an important environmental factor for plants because it changes the day length of each season, and the flowering time of plants is determined by the day length. This is fatal because plants do not disperse seeds if they cannot flower at the appropriate time. This environmental factor may have influenced the historical spread of millets. A comparative genomic analysis suggested that *S. italica* is highly sensitive to temperature and day length (Jia et al. 2013). *S. italica* grown under a non-adaptive photoperiod exhibited a greatly reduced yield (Zhang et al. 2017). Domestication of wild species in the history of human agriculture has involved short term evolutionary changes, primarily driven by microevolution and loss-of-gene function. Unlike genetic divergence, which occurs over millions of years, domestication by humans cannot be expected to be an accumulation of spontaneous de novo mutations. Microevolution entails fixing beneficial genes and eliminating others from the standing variation within the original population. Loss-of-gene function occurs when genes are disrupted, such as via the insertion of transposable elements at functional loci in the genome. Furthermore, because evolution is based on random mutations and events, genes that are not functional in this environment but are useful in other environments may be lost. The rapid spread of millets from one region to a different environment is not easy, because the genetic diversity and functional genes necessary for adaptation to the new region may have already been lost. Genetic diversity can be rescued by crossbreeding with wild species or distantly related cultivated strains. Ancestral wild species of millets may have spread with them as weeds in the agricultural environments, providing genetic diversity to cultivated strains through rare gene flow. Moreover, it is possible that the first propagated strains were not easily established in a new location; therefore, historically, it would be necessary to assume the migration of a large variety of strains and/or multiple transmissions.

History of Proso and Foxtail Millet in Yaponesia

Currently, *P. capillare*, *P. repens*, *P. dichotomiflorum*, and *P. scoparium* are known as naturalized *Panicum* plants in Yaponesia, with *P. bisulcatum* being considered the only Yaponesian native species in the genus *Panicum*. Weedy proso millets, i.e., *P. miliaceum* subsp. *ruderale* and *P. miliaceum* subsp. *agricolum*, have not been found in Yaponesia. Therefore, it is certain that Yaponesian proso millet propagated from the continent. Next, among the *Setaria* species that are closely related to *S. italica*, *S. faberi* is definitely native to Yaponesia but is tetraploid, and *S. pumila*, which has a genome composed of the different subgenomes from *S. italica*, may be native to Yaponesia. However, its diploid wild ancestor, *S. viridis*, is thought to have been introduced into Yaponesia together with *S. italica*. In addition, the genus *Oryza*, including rice, is distributed in the tropics and subtropics. *O. rufipogon* and *O. nivara*, which are closely related to rice (*O. sativa*), have not grown in a wild state in Yaponesia (Stein et al. 2018). These grains were domesticated strains from China that spread to Yaponesia.

Seeds of proso millet, foxtail millet, and wild *Setaria* species have been found in Neolithic pottery impressions in South Korea from approximately 4500 BC (Obata and Manabe 2014). Paddy rice cultivation is considered to have spread to South Korea around 1500 BC. In addition, obsidians from Kyushu, Yaponesia, have been found in South Korea, indicating that exchanges already existed between these areas during the Paleolithic period (Lee and Kim 2015). Thus, *P. miliaceum* and *S. italica* are thought to have been introduced into Yaponesia from South Korea (Fig. 8.5). According to the radiocarbon dating of various samples, including carbonized grains, charcoal, and burned pottery, paddy rice cultivation is thought to have begun in northern Kyushu around 900 BC, at the earliest (Fujio 2011). Furthermore, carbonized grains of foxtail millet embedded in pottery excavated from the Etsuji site in northern Kyushu were found using an X-ray device and were directly dated, which indicated that the seeds dated back to approximately 900 BC (Obata and Kunikita 2022). Northern Kyushu is the site of the oldest paddy rice cultivation in Yaponesia, and many other large-scale agricultural sites from this period have been reported. Moreover, in the same era, 13 bones of ancient native Yaponesian individuals who fed on millet were excavated from the Shimekake site located in Nagano Prefecture, central Honshu, Yaponesia (Yoneda et al. 2021). Impressions of proso millet and foxtail millet found on pottery from this site suggest that native Yaponesian cultivation of millets may have originated from the continent. In addition, more impressions and plant remains of proso millet and foxtail millet compared with those of rice were found at some archeological sites located in the Kanto region around 200 BC (Niiyama 2009; Shitara and Takase 2014). Proso millet and foxtail millet were introduced into Yaponesia after approximately 1000 BC, and then became established throughout Yaponesia. A large green foxtail with cultivated characteristics, is now observed in grasslands and wastelands throughout Yaponesia due to gene flow from *S. italica* into *S. viridis* (Fig. 8.3c) (Osada 1993). The hybrid species is called *S. pycnocoma*. This is thought to be a remnant of the historical spread of foxtail millet cultivation in Yaponesia.

In Hokkaido, the cultivation of proso millet and foxtail millet spread together with that of wheat and barley after the Satsumon Period (700–1300 AD) (Yoshizaki 2003). In the cold regions of Yaponesia, where rice cultivation was difficult, much of the grain production relied on proso and foxtail millet (Yamada 2000). Most of the Yaponesian *P. miliaceum* genotypes were similar to those from the Korean Peninsula; whereas some strains from northern Yaponesia shared a common genotype with strains from northeastern China and Primorsky Krai (Araki et al. 2012). However, there is no archeological evidence that millet was brought to northern Yaponesia from northern China or Primorsky Krai. Nevertheless, because the latitudinal adaptation of millets involved changes at the molecular or microevolutionary level, it is plausible that environmental adaptation yields genes for strains that can grow at the same latitude. The possibility of this another *P. miliaceum* propagation route in northern Yaponesia will need to be investigated in the future.

In addition, proso millet and foxtail millet were important in Okinawa, even after 1400 AD, as they were used as materials for ceremonial liquors on neighboring islands and could be used as a tax item (Kanou 2000). The *S. italica* strain found in

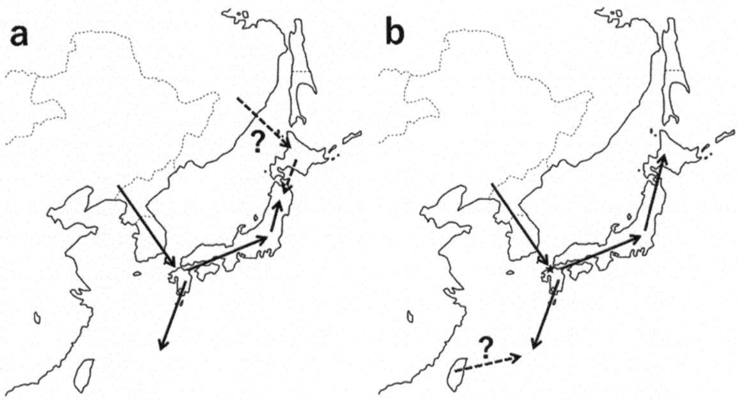

Fig. 8.5 Propagation routes of (**a**) *Panicum miliaceum* and (**b**) *Setaria italica* to Yaponesia. Broken lines indicate pathways supported only by genetic data. The star indicates the site at which the oldest grains of foxtail millet in Yaponesia were excavated

Okinawa shares a common genotype with those from Taiwan and the Philippines (Fukunaga et al. 2002b; Hirano et al. 2011). However, the archeological evidence in support of this phenomenon is also poor. However, because *S. italica* was introduced into Taiwan by sea from Shandong Province, and later migrated from Taiwan to the Philippines, a scenario in which the same strains were also introduced into Okinawa cannot be ruled out. Yaponesia has a long north–south axis, with the southern tip of Okinawa Island at about 26°N, and the northern tip of Hokkaido at about 45.5°N. It would have been difficult to adapt to both the low-latitude subtropical Okinawa and high-latitude subarctic Hokkaido areas based exclusively on the genetic diversity of the millet population that was first introduced to Yaponesia.

8.3 History of Barnyard Millet

From Wild to Domesticated Barnyard Millet

Barnyard millet is a plant of genus *Echinochloa*. Most species of the genus *Echinochloa*, of which approximately 50 have been recorded, including barnyard millet, are distributed from the tropics to the subarctic region, and grow wild in wet or damp areas. Their high phenotypic plasticity regarding the appearance of *Echinochloa* and the morphological similarities across the genus have led to confusion in their taxonomy and species identification (Damalas et al. 2008). *Echinochloa* is an annual or perennial plant that grows to a size of 10–460 cm. The basic chromosome number of the genus *Echinochloa* is x = 9; however, many species exhibit polyploidy. Among *Echinochloa* species, *E. crus-galli* (2n = 6x = 54), *E. colona* (2n = 6x = 54), and *E. oryzicola*, also called *E. phyllopogon* (2n = 4x = 36), have spread worldwide as the most serious weeds in paddy fields. *Echinochloa* weeds generally grow faster and larger than rice in paddy fields. *Echinochloa* species display typical weed characteristics, such as aggressive competition, crop mimicry, allelopathy, high nitrogen utilization, and stress resistance. Because of the widespread use of herbicides to control *Echinochloa* in agricultural environments, resistance to various herbicides evolved among many *Echinochloa* populations worldwide. *Echinochloa* have a strong allelopathic potential, by inhibiting the growth of surrounding plants via the biosynthesis and release of allelochemicals such as benzoxazinone and momilactone, in the rhizosphere. Moreover, *Echinochloa* weeds can reduce rice yields by more than half. Because of these agricultural concerns, research on *Echinochloa* plants has focused primarily on weeding.

E. crus-galli is a native species of East Asia, and its seeds are often found at archeological sites. Moreover, it exhibits high phenotypic plasticity and various genetic phenotypes, resulting in a confusingly large number of defined subspecies. The historically domesticated *E. crus-galli* is *E. esculenta*, also known as Japanese barnyard millet, and is sometimes referred to by another Latin name, *E. utilis* (Fig. 8.6). *E. esculenta* is widely cultivated mainly in East Asia, and is thought to have a genome size of approximately 1.4 Gb because it was domesticated from *E. crus-galli*, which has the same karyotype (Guo et al. 2017; Hilu 1994). The allohexaploid genome of *E. crus-galli* was formed via two hybridization events that occurred during its evolution. One is the hybridization of two diploid *Echinochloa* species approximately 1.9 million years ago, resulting in the allotetraploid *E. oryzicola*, whereas the other is the addition of another subgenome to the allotetraploid genome approximately 0.3 million years ago (Fig. 8.6d) (Ye et al. 2020). The genetic

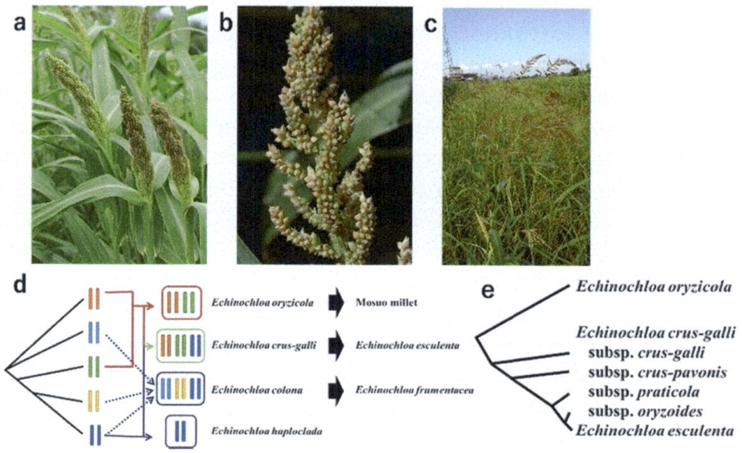

Fig. 8.6 (**a**) and (**b**) *Echinochloa esculenta* (courtesy of National Agriculture and Food Research Organization). (**c**) *Echinochloa crus-galli*. (**d**) Phylogenetic relationships of the subgenomes in the genus *Echinochloa* and the karyotype of each plant. Black arrows indicate domestication. Modified from Ye et al. (2020) and Wu et al. (2022). (**e**) Phylogenetic relationships of subspecies of *E. crus-galli*. Modified from Wu et al. (2022)

diversity of *E. crus-galli* collections from around the world, consisting mainly of wild strains from China, was investigated (Wu et al. 2022). These samples were classified into four weedy subspecies, including subsp. *crus-galli*, subsp. *crus-pavonis*, subsp. *praticola*, and subsp. *oryzoides*, as well as the cultivated *E. esculenta* using a molecular phylogenetic analysis. Two admixed groups were also identified in the *E. crus-galli* subsp. *crus-galli* population. This finding suggests the occurrence of many historical interbreeding events among the subspecies. A molecular phylogenetic analysis of *E. esculenta*, revealed that subsp. *crus-galli* was the most distantly related of the four subspecies, followed by subsp. *crus-pavonis*, subsequently subsp. *praticola* diverged, and subsp. *oryzoides* was the most closely related subspecies (Fig. 8.6e). *E. esculenta* and *E. crus-galli* subsp. *oryzoides* exhibited a lower genetic diversity than subsp. *praticola* ($\pi = 0.00087$, $\pi = 0.00077$ and $\pi = 0.00102$, respectively). Genetic changes in plant tillering structure and seed shattering, which are typical domestication traits of cereals, were observed in subsp. *crus-galli* and subsp. *oryzoides*, and both had larger grain sizes than the remaining subspecies. This evidence indicates that barnyard millet diverged from subsp. *praticola* had been domesticated, and later subsp. *oryzoides* was feralized. In *E. crus-galli*, there are differences among subspecies in terms of distribution area according to latitude, and genes related to flowering time were found to be under natural selection.

E. frumentacea, also known as Indian barnyard millet or sawa millet, is a domesticated cereal that has historically been cultivated in South Asia and sub-Saharan Africa. It has a genome size of approximately 1.2 Gb and was domesticated from *E. colona*, which exhibits the same karyotype (Guo et al. 2017; Hilu 1994). However, *E. frumentacea* displays karyotype variation within the species, with 2n = 36, 54, and 56 strains also reported. *E. colona* has an allohexaploid genome structure that is similar to that of *E. crus-galli*; however two subgenomes are different types from *E. crus-galli*, indicating that it has evolved differently from *E. crus-galli*. Moreover, archeological evidence suggests that its cultivation began in India as early as 5000 BC and in China as early as 3000 BC (Seetharam et al. 1989). *E. frumentacea* also experienced selection on genes involved in plant tillering structure and seed shattering during domestication in parallel with *E. esculenta* (Wu et al. 2022). *E. frumentacea* is traditionally cultivated in South Asia and East Africa. Because *E. frumentacea* is a common crop in South Asia, it has various analogous names in different South Asian languages (Swamy 2023). In East Africa, several undomesticated wild *Echinochloa* grains, such as *E. stagnina*, are also used as food.

In addition, nonshattering barnyard millets, including *E. crus-galli* and *E. oryzicola*, are cultivated by traditional people in several parts of southern China. Although it has now become a weed and often invades paddy fields, there are records of its traditional cultivation by Miaozu (Yamaguchi et al. 1996). The nonshattering trait of barnyard grass seeds is a typical trait in domestication. The nonshattering barnyard millet is still continuously cultivated by Mosuo in Yongning, Yunnan Province. It is called Mosuo millet and is considered that it was domesticated from wild *E. oryzicola*, based on karyotype identity (Yabuno 2001). It has a large size of small ears, lacks of seed shattering and is less innately dormant (Hirosue et al. 2000). It is possible that it has been domesticated in this region because there is no millet in the world that originated from the same wild ancestor. Mosuo millet has green ears, whereas a cultivated nonshattering barnyard millet with taller plants and purple ears is grown in Lijiang, Yunnan, and is called Lijiang millet. A chloroplast DNA sequence analysis of Echinochloa species indicated that Lijiang millet is more closely related to *E. crus-galli* than it is to *E. oryzicola*, suggesting that it was domesticated independently of Mosuo millet (Yamaguchi et al. 2005). Lijiang millet is used by the traditional Naxi for human consumption when grown in rice fields, and for livestock and poultry feed when grown in open fields (Yamaguchi et al. 1996). Cultivated barnyard millet and feralized barnyard grasses are widely reported from Yunnan and Guizhou provinces in southern China. Traditional people in southern China called each subspecies of wild barnyard millets and cultivated barnyard millet by different names, indicating the traditional importance of barnyard millets to them (Yamaguchi et al. 1996). These disappearing Chinese barnyard millets may have historically sustained southern Chinese society as the third domesticated barnyard millet in the world and are worthy of future detailed investigation (Fig. 8.7).

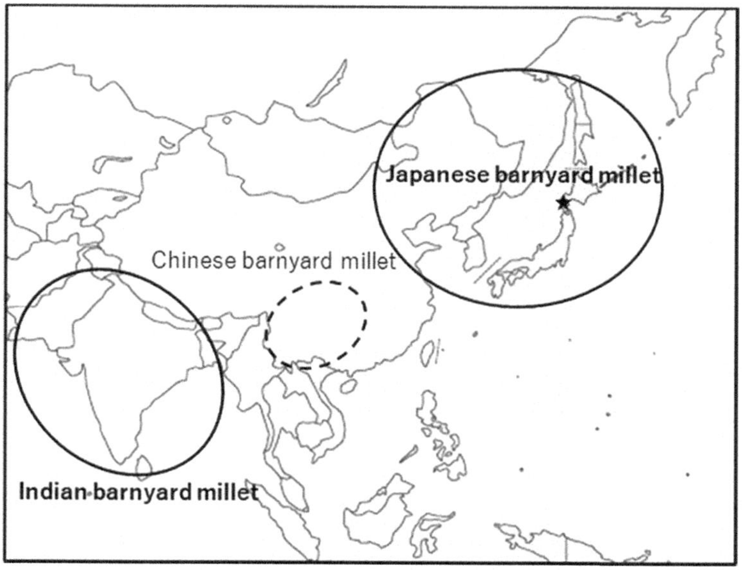

Fig. 8.7 Historical domestication centers of barnyard millets. The broken circle indicates the predicted domestication center of traditional Chinese barnyard millets: i.e., Mosuo millet and Lijiang millet. The star indicates the site where the first grain of domesticated Japanese barnyard millet was excavated

History of Barnyard Millet in East Asia

There is evidence of the use of barnyard millet together with rice in the Yangtze River basin in China at approximately 8000 BC; however, it appears that the millet had not been domesticated (Yang et al. 2015). It is possible that the barnyard millet excavated around the Yangtze River and southern China is related to the origin of the traditional barnyard millets in current southern China; therefore, further research is needed to elucidate this issue. In later times, a few grains of barnyard millet have been found at sites located in the Korean Peninsula dating to before 1500 BC, mixed with rice, proso millet, foxtail millet, wheat, barley, and some legumes (Obata and Manabe 2014). Nevertheless, these could have been mixed with weed seeds growing around the area, or not considered important food. The grains that were cultivated during Neolithic period in South Korea were rice, proso millet, and foxtail millet; Japanese barnyard millet had not been cultivated at that time. *Echinochloa* species have short life cycles, even in the wild, and their seed size sometimes exceeds 2 mm. This size is comparable to that of cultivated proso millet and foxtail millet. For the supplementation of rice, proso millet, and foxtail millet as staple foods, it may have been sufficient to collect barnyard millet in the wild.

Regarding the use of barnyard millet in Yaponesia, carbonized seeds have been excavated from approximately 7000 BC, and plant opals and plant remains since approximately 4000 BC may indicate that Yaponesian millet cultivation had already begun at that time (Yoshizaki 1995; Nasu 2018). Since approximately 3500 BC in northern Yaponesia, the grain size of Japanese barnyard millet has gradually increased over time. Carbonized seeds from 2700 BC, which are round shape and exhibit a larger size compared with the current upper limit observed among wild species, were excavated from sites located in the southmost area of Hokkaido (Crawford 1983). In addition, grains with characteristics of cultivated Japanese barnyard millet have been excavated from many archeological sites in Hokkaido and Aomori, dating from 3500 to 2500 BC (Yoshizaki 1997; Nasu 2018). Therefore, it is considered that *E. esculenta* was domesticated in northern Yaponesia around 2700 BC. According to the prediction based on the number of residential units per site, the population in northern Yaponesia increased rapidly around 3000 BC (Crema et al. 2016). The warmer temperatures occurring in northern Yaponesia around 4000–2000 BC would have contributed to this population increase, as more food was available to support their hunter–gatherer lifestyle (Kawahata 2019). The excavation of the barnyard millet with a large grain size coincides with this period of warming and population increase. In northern Yaponesia, barnyard millet was the only cultivated grain for people who relied on collected acorns and horse chestnuts as alternative carbohydrates, before the introduction of rice, proso millet, and foxtail millet. However, the Jomon sites in northern Yaponesia peaked around 3000 BC, and population genetic analyses suggest that Yaponesian males experienced a significant decrease in population size around 2300 BC (Watanabe et al. 2019). This may be attributed to the return of colder temperatures in northern Yaponesia, which would have reduced the amount of food available through hunting and gathering. Concomitantly, large grains of barnyard millet from subsequent period during a long time have not been found. However, small amounts of barnyard millet not exceeding the current upper limit of grain size for *E. crus-galli* have been continuously excavated after 2300 BC. We do not know whether this is because the sites at which the domesticated millet was maintained have not yet been surveyed, whether the cold weather prevented the millet from growing sufficiently, or whether the genes involved in the production of the large grain size trait were not continuously transmitted in the cultivated barnyard millet. Stable excavations of large barnyard millet grains have been performed at sites from 1000 AD in Yaponesia (Tsubakisaka 2007). It has been suggested that a period of moderate warmth lasted from 850 to 1050 AD in East Asia (Cook et al. 2013), and that the size of the barnyard millet grain on the continent also increased during the same period (Nasu 2018). These increases in millet grain size may have been largely caused by climatic effects. A document from the Nara period (710–794 AD) states that barnyard millet was used to pay taxes in Aichi Prefecture (Yabuno 2001). This means that Japanese barnyard millet had by then acquired a sufficiently important place in Yaponesian culture. Furthermore, in Ainu oral legends, proso millet and foxtail millet were brought from neighboring areas, whereas Japanese barnyard millet was believed to be a direct gift from the gods (Hayashi 1969). It seems certain that the cultivation of *E. esculenta* began in northern Yaponesia; however, the timing of the fixation of the domestication trait remains uncertain. In addition, whether the continental barnyard millet, the grain size of which increased in parallel with that of the barnyard millet in Yaponesia after 850 AD, and the Lijiang millet, which is still cultivated in southern China, have the same origin as *E. esculenta* domesticated in Yaponesia, or were domesticated in parallel, and whether gene flow occurred among those strains, require further investigation.

After the Yayoi period (about 900 BC–300 AD), rice became the most important grain in Yaponesia, because of the widespread implementation of paddy rice cultivation. *E. esculenta* had low waxiness and was less suitable than rice, proso millet, and foxtail millet for steaming and mochi processing. The waxy grain of *E. esculenta* was artificially introduced during modern breeding and has not been found in landraces (Hoshino et al. 2010). This may be attributed to the difficulty in deleting all the *Waxy* genes present in the hexaploid genome. Perhaps for this reason, it was sometimes said that Japanese barnyard millet was inferior to rice in taste. However, one of the main reasons that Japanese barnyard millet has continued to be cultivated is that it is more resistant to cold damage than rice. Therefore, millet cultivation, including that of *E. esculenta*, had been important in the mountainous regions and cold region. In particular, in Hokkaido and Primorsky Krai, the cultivation of barnyard millet continued even after 700 AD (Yamada 2000). The alcohol used during the Iomante, which is an important Ainu ritual, was called tonoto and brewed from Japanese barnyard millet (Hayashi 1969). In fact, Japanese barnyard millet was traditionally brewed in many parts of Yaponesia, and Mosuo millet is brewed as Sulima by the Mosuo people of Yunnan Province, China. Barnyard millet is popular for brewing alcohol and has historically been cultivated as food mainly in the cooler regions of East Asia, where its yields is more consistent than rice even during cool summers.

8.4 Comments

In addition to rice, proso millet, foxtail millet, and barnyard millet have historically been important grains in Yaponesian and East–Asian agriculture. In these regions, which include the northern limit of rice cultivation, these millets, which can be grown at high latitudes and are resistant to cold, must have been extremely valuable. Because of recent global warming and the occurrence of abnormal weather, millets that can be grown in hot and arid regions are attracting attention; however, grains that can be grown at low temperatures and high latitudes will also be valuable. Millet research will increasingly focus on elucidating the history of agriculture in Yaponesia, East Asia, and northern Eurasia and guiding the future of agriculture.

References

Abbo S, Pinhasi van-Oss R, Gopher A, Saranga Y, Ofner I, Peleg Z (2014) Plant domestication versus crop evolution: a conceptual framework for cereals and grain legumes. Trends Plant Sci 19:351–360

Araki M, Numaoka A, Kawase M, Fukunaga K (2012) Origin of waxy common millet, *Panicum miliaceum* L. in Japan. Genet Resour Crop Evol 59:1303–1308

Barton L, Newsome SD, Chen F-H, Wang H, Guilderson TP, Bettinger RL (2009) Agricultural origins and the isotopic identity of domestication in northern China. Proc Natl Acad Sci U S A 106:5523–5528

Bennetzen JL, Schmutz J, Wang H, Percifield R, Hawkins J, Pontaroli AC, Estep M, Feng L, Vaughn JN, Grimwood J, Jenkins J, Barry K, Lindquist E, Hellsten U, Deshpande S, Wang X, Wu X, Mitros T, Triplett J, Yang X, Ye C-Y, Mauro-Herrera M, Wang L, Li P, Sharma M, Sharma R, Ronald PC, Panaud O, Kellogg EA, Brutnell TP, Doust AN, Tuskan GA, Rokhsar D, Devos KM (2012) Reference genome sequence of the model plant Setaria. Nat Biotechnol 30:555–561

Chen J, Liu Y, Liu M, Guo W, Wang Y, He Q, Chen W, Liao Y, Zhang W, Gao Y, Dong K, Ren R, Yang T, Zhang L, Qi M, Li Z, Zhao M, Wang H, Wang J, Qiao Z, Li H, Jiang Y, Liu G, Song X, Deng Y, Li H, Yan F, Dong Y, Li Q, Li T, Yang W, Cui J, Wang H, Zhou Y, Zhang X, Jia G, Lu P, Zhi H, Tang S, Diao X (2023) Pangenome analysis reveals genomic variations associated with domestication traits in broomcorn millet. Nat Genet 55:2243–2254

Cook ER, Krusic PJ, Anchukaitis KJ, Buckley BM, Nakatsuka T, Sano M (2013) Tree-ring reconstructed summer temperature anomalies for temperate East Asia since 800 C.E. Clim Dyn 41:2957–2972

Crawford WG (1983) Paleoethnobotany of the Kameda Peninsula Jomon. University of Michigan Press, Ann Arbor, MI

Crema ER, Habu J, Kobayashi K, Madella M (2016) Summed probability distribution of 14C dates suggests regional divergences in the population dynamics of the Jomon period in Eastern Japan. PLoS One 11:e0154809

Damalas CA, Dhima KV, Eleftherohorinos IG (2008) Morphological and physiological variation among species of the genus *Echinochloa* in Northern Greece. Weed Sci 56:416–423

de Wet JMJ, Brink DE, Rao KEP, Mengesha MH (1983) Diversity in kodo millet, *Paspalum scrobiculatum*. Econ Bot 37:159–163

Dong G, Yang Y, Han J, Wang H, Chen F (2017) Exploring the history of cultural exchange in prehistoric Eurasia from the perspectives of crop diffusion and consumption. Sci China Earth Sci 60:1110–1123

Doust AN, Kellogg EA (2006) Effect of genotype and environment on branching in weedy green millet (*Setaria viridis*) and domesticated foxtail millet (*Setaria italica*) (Poaceae). Mol Ecol 15:1335–1349

Fujio S'i (2011) The 'new' Yayoi period: paddy field rice cultivation five hundred years earlier. Yoshikawa Kobunkan, Tokyo. (in Japanese)

Fukunaga K, Kawase M, Kato K (2002a) Structural variation in the *Waxy* gene and differentiation in foxtail millet [*Setaria italica* (L.) P. Beauv.]: implications for multiple origins of the waxy phenotype. Mol Gen Genomics 268:214–222

Fukunaga K, Wang Z, Kato K, Kawase M (2002b) Geographical variation of nuclear genome RFLPs and genetic differentiation in foxtail millet, *Setaria italica* (L.) P. Beauv. Genet Resour Crop Evol 49:95–101

Fuller DQ (2007) Contrasting patterns in crop domestication and domestication rates: recent archaeobotanical insights from the Old World. Ann Bot 100:903–924

Guo L, Qiu J, Ye C, Jin G, Mao L, Zhang H, Yang X, Peng Q, Wang Y, Jia L, Lin Z, Li G, Fu F, Liu C, Chen L, Shen E, Wang W, Chu Q, Wu D, Wu S, Xia C, Zhang Y, Zhou X, Wang L, Wu L, Song W, Wang Y, Shu Q, Aoki D, Yumoto E, Yokota T, Miyamoto K, Okada K, Kim D-S, Cai D, Zhang C, Lou Y, Qian Q, Yamaguchi H, Yamane H, Kong C-H, Timko MP, Bai L, Fan L (2017) *Echinochloa crus-galli* genome analysis provides insight into its adaptation and invasiveness as a weed. Nat Commun 8:1031

Hamoud MA, Haroun SA, MacLeod RD, Richards AJ (1994) Cytological relationships of selected species of *Panicum* L. Biol Plant 36:37–45

Hayashi Y (1969) The Ainu agriculture. Keiyuusya, Tokoyo. (in Japanese)

He K, Lu H, Zhang J, Wang C, Huan X (2017) Prehistoric evolution of the dualistic structure mixed rice and millet farming in China. The Holocene 27:1885–1898

Hilu KW (1994) Evidence from RAPD markers in the evolution of *Echinochloa* millets (*Poaceae*). Plant Syst Evol 189:247–257

Hirano R, Naito K, Fukunaga K, Watanabe KN, Ohsawa R, Kawase M (2011) Genetic structure of landraces in foxtail millet (*Setaria italica* (L.) P. Beauv.) revealed with transposon display and interpretation to crop evolution of foxtail millet. Genome 54:498–506

Hiremath SC, Patil GNV, Salimath SS (1990) Genome homology and origin of *Panicum sumatrense* (Gramineae). Cytologia (Tokyo) 55:315–319

Hirosue T, Yamasue Y, Yabuno T (2000) Shattering habit and dormancy of spikelets in a cultivated form of *Echinochloa oryzicola* recently found in China. Weed Res 40:449–456

Hoshino T, Nakamura T, Seimiya Y, Kamada T, Ishikawa G, Ogasawara A, Sagawa S, Saito M, Shimizu H, Nishi M, Watanabe M, Takeda J, Takahata Y (2010) Production of a fully waxy line and analysis of waxy genes in the allohexaploid crop, Japanese barnyard millet. Plant Breed 129:349–355

Hung H-C (2005) Neolithic interaction between Taiwan and Northern Luzon: the pottery and jade evidences from the Cagayan valley. J Austronesian Stud 1:109–133

Hunt HV, Badakshi F, Romanova O, Howe CJ, Jones MK, Heslop-Harrison JSP (2014) Reticulate evolution in *Panicum* (Poaceae): the origin of tetraploid broomcorn millet, *P. miliaceum*. J Exp Bot 65:3165–3175

Jia G, Huang X, Zhi H, Zhao Y, Zhao Q, Li W, Chai Y, Yang L, Liu K, Lu H, Zhu C, Lu Y, Zhou C, Fan D, Weng Q, Guo Y, Huang T, Zhang L, Lu T, Feng Q, Hao H, Liu H, Lu P, Zhang N, Li Y, Guo E, Wang S, Wang S, Liu J, Zhang W, Chen G, Zhang B, Li W, Wang Y, Li H, Zhao B, Li J, Diao X, Han B (2013) A haplotype map of genomic variations and genome-wide association studies of agronomic traits in foxtail millet (*Setaria italica*). Nat Genet 45:957–961

Kanou A (2000) Revival and background of the traditional crop proso millet in Tonaki and Aguni Islands, Okinawa, Japan. Hum Geogr 52:67–83. (in Japanese)

Kawahata H (2019) Climatic reconstruction at the Sannai-Maruyama site between bond events 4 and 3—implication for the collapse of the society at 4.2 ka event. Prog Earth Planet Sci 6. https://doi.org/10.1186/s40645-019-0308-8

Kihara H, Kishimoto E (1942) Hybrids between *Setalia italica* and *Setaria viridis*. Bot Mag 56:62–67. (in Japanese)
Kitagawa M (1937) A contribution to the Flora of Manchuria. Bot Mag Tokyo 51:153–154
Lee G, Kim J (2015) Obsidians from the Sinbuk archaeological site in Korea—evidences for strait crossing and long-distance exchange of raw material in Paleolithic age. J Archaeol Sci Rep 2:458–466
Li Y (2021) Agriculture and palaeoeconomy in prehistoric Xinjiang, China (3000–200 BC). Veget Hist Archaeobot 30:287–303
Li HW, Li CH, Pao WK (1945) Cytological and genetical studies of the interspecific cross of the cultivated foxtail millet, *Setaria italica (L.)* Beauv., and the green foxtail millet, *S. viridis L.* Agron J 37:32–54
Li S-X, Wang Z-H, Malhi SS, Li S-Q, Gao Y-J, Tian X-H (2009) Nutrient and water management effects on crop production, and nutrient and water use efficiency in dryland areas of China (Chapter 7). In: Advances in agronomy, vol 102. Elsevier, pp 223–265
Li C, Liu M, Sun F, Zhao X, He M, Li T, Lu P, Xu Y (2021) Genetic divergence and population structure in weedy and cultivated broomcorn millets (*Panicum miliaceum L.*) revealed by specific-locus amplified fragment sequencing (SLAF-seq). Front Plant Sci 12:688444
Li X, Gao J, Song J, Guo K, Hou S, Wang X, He Q, Zhang Y, Zhang Y, Yang Y, Tang J, Wang H, Persson S, Huang M, Xu L, Zhong L, Li D, Liu Y, Wu H, Diao X, Chen P, Wang X, Han Y (2022) Multi-omics analyses of 398 foxtail millet accessions reveal genomic regions associated with domestication, metabolite traits, and anti-inflammatory effects. Mol Plant 15:1367–1383
Lisitsyna GN, Prishchepenko LV (1977) Palaeoethnobotanical finds of the Caucasus and the near east. Nauka, Moscow. (in Russian)
Lu H, Zhang J, Liu K-B, Wu N, Li Y, Zhou K, Ye M, Zhang T, Zhang H, Yang X, Shen L, Xu D, Li Q (2009) Earliest domestication of common millet (*Panicum miliaceum*) in East Asia extended to 10, 000 years ago. Proc Natl Acad Sci U S A 106:7367–7372
Majure LC (2009) *Panicum repens* (torpedo grass) in Mississippi. J Miss Acad Sci 54:164–171
Mamidi S, Healey A, Huang P, Grimwood J, Jenkins J, Barry K, Sreedasyam A, Shu S, Lovell JT, Feldman M, Wu J, Yu Y, Chen C, Johnson J, Sakakibara H, Kiba T, Sakurai T, Tavares R, Nusinow DA, Baxter I, Schmutz J, Brutnell TP, Kellogg EA (2020) A genome resource for green millet *Setaria viridis* enables discovery of agronomically valuable loci. Nat Biotechnol 38:1203–1210
Martin L, Messager E, Bedianashvili G, Rusishvili N, Lebedeva E, Longford C, Hovsepyan R, Bitadze L, Chkadua M, Vanishvili N, Le Mort F, Kakhiani K, Abramishvili M, Gogochuri G, Murvanidze B, Giunashvili G, Licheli V, Salavert A, Andre G, Herrscher E (2021) The place of millet in food globalization during Late Prehistory as evidenced by new bioarchaeological data from the Caucasus. Sci Rep 11:13124
Masterson J (1994) Stomatal size in fossil plants: evidence for polyploidy in majority of angiosperms. Science 264:421–424
Nabhan G, de Wet JMJ (1984) *Panicum sonorum* in Sonoran Desert agriculture. Econ Bot 38:65–82
Nasu H (2018) Domestrication of plants during the Jomon period. Quat Res 57:109–126. (in Japanese)
Niiyama M (2009) Identification of carbonized seeds excavated from clay pit. In: Nakayashiki Site Excavation Team of the Department History and Culture the Faculty of Humanities and Culture Showa Women's University (ed) (ed) Nakayashiki site excavation report. The Department History and Culture the Faculty of Humanities and Culture Showa Women's University, pp 145–147. (in Japanese)
Obata H, Kunikita D (2022) A new archaeological method to reveal the arrival of cereal farming: development of a new method to extract and date of carbonised material in pottery and its application to the Japanese archaeological context. J Archaeol Sci 143:105594
Obata H, Manabe A (2014) Pottery impressions and early farming in the Korean Jeulmun pottery culture. Bull Natl Museum Jpn Hist 187. (in Japanese)
Osada T (1993) Illustrated grasses of Japan. Heibonsha, Tokyo. (in Japanese)

Robbeets M, Bouckaert R, Conte M, Savelyev A, Li T, An D-I, Shinoda K-I, Cui Y, Kawashima T, Kim G, Uchiyama J, Dolińska J, Oskolskaya S, Yamano K-Y, Seguchi N, Tomita H, Takamiya H, Kanzawa-Kiriyama H, Oota H, Ishida H, Kimura R, Sato T, Kim J-H, Deng B, Bjørn R, Rhee S, Ahn K-D, Gruntov I, Mazo O, Bentley JR, Fernandes R, Roberts P, Bausch IR, Gilaizeau L, Yoneda M, Kugai M, Bianco RA, Zhang F, Himmel M, Hudson MJ, Ning C (2021) Triangulation supports agricultural spread of the Transeurasian languages. Nature 599:616–621

Sagart L, Hsu T-F, Tsai Y-C, Wu C-C, Huang L-T, Chen Y-C, Chen Y-F, Tseng Y-C, Lin H-Y, Hsing Y-IC (2018) A northern Chinese origin of Austronesian agriculture: new evidence on traditional Formosan cereals. Rice (N Y) 11:57

Scholz H (1983) Die Unkraut-Hirse (*Panicum miliaceum* subsp. *ruderale*)—neue Tatsachen und Befunde. Plant Syst Evol 143:233–244

Seetharam A, Riley KW, Harinarayana G (eds) (1989) Small millets in global agriculture. Intercept, Wimborne

Shi J, Ma X, Zhang J, Zhou Y, Liu M, Huang L, Sun S, Zhang X, Gao X, Zhan W, Li P, Wang L, Lu P, Zhao H, Song W, Lai J (2019) Chromosome conformation capture resolved near complete genome assembly of broomcorn millet. Nat Commun 10:464

Shitara H, Takase K (2014) The beginning of the cereal cultivation in the South-West Kanto District by the analysis of replica. Bull Natl Museum Jpn Hist 185:511–530. (in Japanese)

Stein JC, Yu Y, Copetti D, Zwickl DJ, Zhang L, Zhang C, Chougule K, Gao D, Iwata A, Goicoechea JL, Wei S, Wang J, Liao Y, Wang M, Jacquemin J, Becker C, Kudrna D, Zhang J, Londono CEM, Song X, Lee S, Sanchez P, Zuccolo A, Ammiraju JSS, Talag J, Danowitz A, Rivera LF, Gschwend AR, Noutsos C, Wu C-C, Kao S-M, Zeng J-W, Wei F-J, Zhao Q, Feng Q, El Baidouri M, Carpentier M-C, Lasserre E, Cooke R, da Rosa Farias D, da Maia LC, dos Santos RS, Nyberg KG, McNally KL, Mauleon R, Alexandrov N, Schmutz J, Flowers D, Fan C, Weigel D, Jena KK, Wicker T, Chen M, Han B, Henry R, Hsing Y-IC, Kurata N, de Oliveira AC, Panaud O, Jackson SA, Machado CA, Sanderson MJ, Long M, Ware D, Wing RA (2018) Genomes of 13 domesticated and wild rice relatives highlight genetic conservation, turnover and innovation across the genus *Oryza*. Nat Genet 50:285–296

Stevens CJ, Fuller DQ (2017) The spread of agriculture in eastern Asia. Lang Dyn Change 7:152–186

Stevens CJ, Murphy C, Roberts R, Lucas L, Silva F, Fuller DQ (2016) Between China and South Asia: a Middle Asian corridor of crop dispersal and agricultural innovation in the Bronze Age. Holocene 26:1541–1555

Stevens CJ, Shelach-Lavi G, Zhang H, Teng M, Fuller DQ (2021) A model for the domestication of *Panicum miliaceum* (common, proso or broomcorn millet) in China. Veget Hist Archaeobot 30:21–33

Stoltenberg DE, Wiederholt RJ (1995) Giant foxtail (*Setaria faberi*) resistance to aryloxyphenoxy-propionate and cyclohexanedione herbicides. Weed Sci 43:527–535

Swamy KRM (2023) Origin, domestication, taxonomy, botanical description, genetics and cytogenetics, genetic diversity and breeding of barnyard millet (*Echinochloa frumentacea Link.*). Int J Curr Res Rev 15:25839–25864

Till-Bottraud I, Reboud X, Brabant P, Lefranc M, Rherissi B, Vedel F, Darmency H (1992) Outcrossing and hybridization in wild and cultivated foxtail millets: consequences for the release of transgenic crops. Theor Appl Genet 83:940–946

Triplett JK, Wang Y, Zhong J, Kellogg EA (2012) Five nuclear loci resolve the polyploid history of switchgrass (*Panicum virgatum L.*) and relatives. PLoS One 7:e38702

Tsubakisaka Y (2007) Echinochloa plants excavated from archaeological sites. In: Yamaguchi H (ed) Archaeological and ethnobotanical studies on Ainu barnyard millet wine. Report of research funded by the organization for the promotion of Ainu culture and research. Osaka Metropolitan University, pp 14–25. (in Japanese)

Wang RL, Dekker JH (1995) Weedy adaptation in Setaria spp.: III. Variation in herbicide resistance in Setaria spp. Pestic Biochem Physiol 51:99–116

Wang C, Chen J, Zhi H, Yang L, Li W, Wang Y, Li H, Zhao B, Chen M, Diao X (2010) Population genetics of foxtail millet and its wild ancestor. BMC Genet 11:90

Watanabe Y, Naka I, Khor S-S, Sawai H, Hitomi Y, Tokunaga K, Ohashi J (2019) Analysis of whole Y-chromosome sequences reveals the Japanese population history in the Jomon period. Sci Rep 9:8556

Weber S, Kashyap A (2016) The vanishing millets of the Indus civilization. Archaeol Anthropol Sci 8:9–15

Wu D, Shen E, Jiang B, Feng Y, Tang W, Lao S, Jia L, Lin H-Y, Xie L, Weng X, Dong C, Qian Q, Lin F, Xu H, Lu H, Cutti L, Chen H, Deng S, Guo L, Chuah T-S, Song B-K, Scarabel L, Qiu J, Zhu Q-H, Yu Q, Timko MP, Yamaguchi H, Merotto A Jr, Qiu Y, Olsen KM, Fan L, Ye C-Y (2022) Genomic insights into the evolution of *Echinochloa* species as weed and orphan crop. Nat Commun 13:689

Yabuno T (2001) Phylogeny and taxonomy in genus *Echinochloa*. In: Yabuno T, Yamaguchi H (eds) Natural history of genus *Echinochloa*. Zennoukyou, Tokyo, pp 15–30. (in Japanese)

Yamada G (2000) Cultivation of the Satsumon culture. In: Journal of the Hokkaido Archaeological Society. Historical Museum of Hokkaido, pp 15–25

Yamaguchi H, Umemoto S, Masanaga Y (1996) Studies on barnyard grasses, especially on non-shattering form of *Echinochloa oryzicola* Vasing., in Yunkui Plateau, China. Weed Res Jpn 41:111–115

Yamaguchi H, Utano A, Yasuda K, Yano A, Soejima A (2005) A molecular phylogeny of wild and cultivated *Echinochloa* in East Asia inferred from non-coding region sequences of *trn* T-L-F. Weed Biol Manag 5:210–218

Yang X, Fuller DQ, Huan X, Perry L, Li Q, Li Z, Zhang J, Ma Z, Zhuang Y, Jiang L, Ge Y, Lu H (2015) Barnyard grasses were processed with rice around 10000 years ago. Sci Rep 5:16251

Ye C-Y, Wu D, Mao L, Jia L, Qiu J, Lao S, Chen M, Jiang B, Tang W, Peng Q, Pan L, Wang L, Feng X, Guo L, Zhang C, Kellogg EA, Olsen KM, Bai L, Fan L (2020) The genomes of the allohexaploid *Echinochloa crus-galli* and its progenitors provide insights into polyploidization-driven adaptation. Mol Plant 13:1298–1310

Yoneda M, Nakazawa M, Tanaka K, Takahashi Y (2021) Low-level food production with millet cultivation in the Late Jomon period as indicated by isotope analysis of human bones excavated from the Shimekake site, Nagano Prefecture, Japan. J Jpn Archaeol Assoc 53:25–40. (in Japanese)

Yoshizaki M (1995) Emergence of cultivated plant species in Japan. Archaeol Q 50:18–24. (in Japanese)

Yoshizaki M (1997) Domesticated plants of the Jomon period. Quat Res 36:343–346. (in Japanese)

Yoshizaki M (2003) Prehistoric millets: archaeobotany of Japanese barnyard millet and azuki beans. In: Yamaguchi H, Kawase M (eds) Natural history of millets—in search of its origins and culture. Hokkaido University Press, pp 52–70. (in Japanese)

Zeder MA (2015) Core questions in domestication research. Proc Natl Acad Sci U S A 112:3191–3198

Zhang G, Liu X, Quan Z, Cheng S, Xu X, Pan S, Xie M, Zeng P, Yue Z, Wang W, Tao Y, Bian C, Han C, Xia Q, Peng X, Cao R, Yang X, Zhan D, Hu J, Zhang Y, Li H, Li H, Li N, Wang J, Wang C, Wang R, Guo T, Cai Y, Liu C, Xiang H, Shi Q, Huang P, Chen Q, Li Y, Wang J, Zhao Z, Wang J (2012) Genome sequence of foxtail millet (*Setaria italica*) provides insights into grass evolution and biofuel potential. Nat Biotechnol 30:549–554

Zhang K, Fan G, Zhang X, Zhao F, Wei W, Du G, Feng X, Wang X, Wang F, Song G, Zou H, Zhang X, Li S, Ni X, Zhang G, Zhao Z (2017) Identification of QTLs for 14 agronomically important traits in Setaria italica based on SNPs generated from high-throughput sequencing. G3 (Bethesda) 7:1587–1594

Chapter 9
Genomic Studies on the Origin, Hybridization and Dispersal of Asian Rice

Masahiko Kumagai and Hiroaki Sakai

Abstract The cultivated rice, *Oryza sativa*, originating from Asia, is one of the most important cereals and has been extensively studied over several decades. It has the smallest genome size among major cereals and was the first to construct a complete reference genome. Furthermore, genetic polymorphism data have been analyzed in numerous studies, focusing on the geographical origin and relationships between cultivated and wild species. The classification of Asian rice into two subspecies, *Oryza sativa* L. ssp. *japonica* and ssp. *indica*, is well established, and the elucidation of their genomic backgrounds has revealed intricate relationships between these two varietal groups. Despite advancements in the amount of data and analytical techniques, the mysteries surrounding the origin and domestication processes remain subjects of ongoing debate owing to the complexity of the genomic backgrounds. Moreover, several studies utilizing genomic data from over 1000 cultivated and wild rice samples, represented by the 3000 Rice Genomes Project, have initiated discussions regarding the dispersal history of regional populations, transitions of genomic backgrounds during domestication, and genome-wide scans for selection signatures. This paper provides an overview of the classification of rice varietal groups, controversies surrounding the origin and domestication processes of *japonica* and *indica*, and recent findings on population histories revealed by population genomic analyses.

Keywords *Oryza sativa* · Genomics · Diversity · Population · Dispersal

M. Kumagai (✉) · H. Sakai
Research Center for Advanced Analysis, National Agriculture and Food Research Organization, Tsukuba, Ibaraki, Japan
e-mail: kumagai.m@naro.affrc.go.jp; hirsakai@affrc.go.jp

N. Osada et al. (eds.), *Phylogeographic History of Plants and Animals Coexisting with Humans in Asia*, Evolutionary Studies,
https://doi.org/10.1007/978-981-97-6887-5_9

Abbreviations

BAC	Bacterial artificial chromosome
IRGSP	International Rice Genome Sequencing Project
JRC	Japanese Rice Collection
NARO	National Agriculture and Food Research Organization
NGS	Next-generation sequencing
PAC	P1-derived artificial chromosome
SNP	Single Nucleotide Polymorphisms
WRC	World Rice Collection

9.1 Introduction

Domestication of crops and animals began approximately 10,000 years ago, drastically changing the lifestyle and culture of humans, who until then had been primarily hunter-gatherers. This change was caused by the effects of global climate change, which has occurred simultaneously in various regions of the world. The use of cultivated plants and livestock species has continued from ancient times to the present, interacting with humans and the environment and changing the genetic background. The improvement of these varieties is still being conducted as modern breeding, and is an important area of research in agricultural science. For example, improved eating quality can enrich our diet, and increased yields are expected to contribute to food shortage problems associated with global population increase. In recent years, a pollen allergy-relieving rice variety has been developed to prevent hay fever, adding value not only as a source of nutrition, but also as a functional food. Asian cultivated rice, *Oryza sativa*, is the most important cereal, feeding more than 1/3 of the world's population. Asian cultivated rice can be broadly classified into *japonica* and *indica*; however, there are several additional groups (varietal groups or sub-varieties) that are adapted to diverse growing environments, from tropical to temperate regions, and under a variety of moisture conditions (Morishima et al. 1992). How have humans created such a variety of rice cultivars since the beginning of cultivation, and what kinds of rice plants have been used by humans at various times in antiquity? Modern large-scale genomic data obtained by next-generation sequencing (NGS) technologies have enabled us to understand comprehensive genetic diversity, relationships between populations, and the history of rice diffusion. Here, we aim to introduce the current knowledge on rice domestication history, its origin, and diffusion into Asia, from genomic data as well as what pre-genomic studies have revealed.

9.2 Asian Rice *Oryza sativa* and Wild Relatives

The cultivated rice, *Oryza sativa*, belongs to the genus *Oryza* of Poaceae, which is a large family of angiosperm monocotyledons that is widely distributed throughout the world and is commonly known as grasses. Genus *Oryza* is comprised with two cultivated rice, Asian rice *O. sativa* and African rice *O. glaberrima*, and 22 wild relatives (Table 9.1). These wild rice species have a wide range of geographical distribution in pan-tropical regions, including Asia, Australia, Oceania, Africa, and Central and South America (Vaughan 1994; Khush 1997), and exhibit highly diverse morphological features and adaptations to different

Table 9.1 Genus *Oryza* species (Oryzabase n.d.)

Species	Genome	Geographical distribution
Oryza sativa complex		
O. sativa	AA	All over the world
O. rufipogon	AA	Asia, Oceania
O. nivara (annual ecotype of *O. rufipogon*)	AA	Asia, Oceania
O. glaberrima	AA	West Africa
O. barthii	AA	Africa
O. longistaminata	AA	Africa
O. meridionalis	AA	Australia
O. glumaepatula	AA	Central and South America
O. officinalis complex		
O. officinalis	CC	Asia
O. minuta	BBCC	Philippines
O. rhizomatis	CC	Sri Lanka
O. eichingeri	CC	Africa, Sri Lanka
O. punctata	BB, BBCC	Africa
O. latifolia	CCDD	Central and South America
O. alta	CCDD	Central and South America
O. grandiglumis	CCDD	South America
O. australiensis	EE	Australia
O. ridleyi complex		
O. ridleyi	HHJJ	Asia, New Guinea
O. longiglumis	HHJJ	New Guinea
O. granulata complex		
O. granulata	GG	Asia
O. meyeriana	GG	Asia
Others		
O. brachyantha	FF	Africa
O. schlechteri	Unknown	New Guinea
O. coarctata (also known as *Porteresia coarctata*)	HHKK	Coastal region of India, Pakistan and Bangladesh

habitats, as well as the ability to resist biotic and abiotic stresses. Wild *Oryza* species have 15 million years of evolutionary diversification and diverse genomes. There are diploid (2n = 24) and polyploid (2n = 48) genomes, represented as the AA, BB, CC, BBCC, CCDD, EE, FF, GG, HHJJ, and HHKK genome types (Li 1964; Morinaga 1964; Aggarwal et al. 1997) with genome sizes ranging from approximately 300 Mb to 1.2 Gb. From phylogenetic studies, the genus *Oryza* is divided into five genetically related groups: the *O. sativa* complex (AA), the *O. officinalis* complex (BB, CC, EE, BBCC, and CCDD), the *O. ridleyi* complex (HHJJ), the *O. granulata* complex (GG), and others (FF and HHKK) (Wang et al. 1992; Aggarwal et al. 1999; Ge et al. 1999; Joshi et al. 2000; Mullins and Hilu 2002; Nishikawa et al. 2005; Zou et al. 2008; Kumagai et al. 2010). The *O. sativa* complex and the *O. officinalis* complex are large groups and both of them covers entire distribution ranges of genus *Oryza* in Asia, Africa and Americas. Whereas the *O. ridleyi* complex is distributed in Asia and New Guinea and the *O. granulata* complex is distributed only in Asia. The overlapping distribution of the diversified species complex indicated repeated global dispersal of this genus.

Cultivated and wild rice species of the genus *Oryza* are among the most widely studied plant species in terms of genomics. The high-quality complete genome sequence of the *O. sativa japonica* cultivar Nipponbare was constructed by Sanger sequencing based on a physical map of the P1-derived artificial chromosome (PAC) and bacterial artificial chromosome (BAC) by the International Rice Genome Sequencing Project (IRGSP) (International Rice Genome Sequencing Project 2005). IRGSP-1.0, an improved version with error correction using NGS technologies (Kawahara et al. 2013), serves as the basis for cereal genomics studies (Matsumoto et al. 2016). For *indica*, several genomes have been assembled de novo using NGS technologies (Sakai et al. 2014; Schatz et al. 2014; Du et al. 2017). Also, African domesticated rice species *O. glaberrima* genome was sequenced by whole genome shotgun and distinct evolutionary patterns were observed (Sakai et al. 2011). The International *Oryza* Map Alignment Project (I-OMAP) has created a comparative genomics platform for *Oryza* species, including wild rice genomes. Reference genomes for six wild *Oryza* species, *O. rufipogon*, *O. nivara*, *O. barthii*, *O. glumaepatula, O. meridionalis*, *O. punctata* and a relative species *Leersia perrieri* with two domesticated rice *O. sativa*, *indica* IR8 (known as Miracle Rice) and *aus* N22 (drought tolerant rice) were created by using long read and short read NGS with Chromosome level quality (Stein et al. 2018). This study comprehensively compared the genomes of rice species and found that lineage-specific emergence and turnover of genetic elements, including transposons and coding and noncoding genes, contribute to the diversification of *Oryza* species. Recently, 12 high-quality near-gap-free reference sequences representing a subpopulation of *O. sativa* have been published (Zhou et al. 2020). Together with IRGSP-1.0 (Kawahara et al. 2013) and three already published *indica* genomes (Zhang et al. 2016a, b; Stein et al. 2018), these platinum standard reference genomes serve as a resource for pan-genome analysis of *Oryza* species (Zhou et al. 2020).

9.3 Classification of Varietal Groups of *Oryza sativa* Based on Morphology, Hybrid Fertility and Isozyme Analysis

Cultivated rice, *O. sativa*, is largely divided into two main varietal groups, *japonica* and *indica*, scientifically named by Kato et al. (1928) based on classification by studying a combination of morphological features such as shape and color of leaf, angle of top leaf and stem, grain shape, presence and absence of awn and awn length if precent, and hybrid fertility between varieties. This study investigated 90 representative rice varieties collected worldwide (Japan, Korea, Taiwan, China, Indonesia, India, Sri Lanka, Hawaii, Brazil, and the USA), including the Chinese varietal groups Keng and Hsien, which had been empirically distinguished since ancient times. A third type of varietal group, later referred to as *javanica* based on geographical distribution, was recognized on a morphological basis (Matsuo 1952; Morinaga 1954). Later, *japonica* and *javanica* were pointed out to be considerable as temperate and tropical components of a single *japonica* group (Oka 1958). An isozyme study utilizing a large number of varieties was carried out, and six groups, namely groups I to VI, of *O. sativa* were identified (Glaszmann 1987). This study surveyed 15 polymorphic loci coding for eight enzymes for 1688 traditional Asian rice varieties including varietal groups having well-known local names or ecotypes such as Aman, Aus and Boro from Bangladesh, Basmati from Pakistan, Bulu, Gundil and Tjereh from Indonesia (Java and Bali), Hsien and Keng from China, and upland and lowland rice. The classification of the two major groups, *japonica* and *indica*, was supported, and additional four groups were recognized. Group I consisted of *indica*, presented in whole tropical area and dominant in Indian subcontinent, including Aman, Tjereh, Hsien and lowland rice from Thailand and Laos. Group VI consisted of *japonica* and *javanica* rices, from temperate areas and high elevation areas in Southeast Asia and South Asia, including Bulu, Gundil, Keng, upland rice from Thailand and Laos, and Korean and Japanese varieties. Group II was exclusively observed in South and West Asia, and was frequent along the Himalayas, from Iran to Assam hills, and corresponded to Aus and Boro. Groups III and IV were minor groups corresponding to the local varietal groups and adapted to deep-water conditions. Group III was found only in Bangladesh and the Manipur State of India. Group IV consisted of Rayada from Bangladesh. Group V consisted of very diverse varieties from broader areas, including high-quality rice varieties, such as Sadri from Iran and Basmati from Pakistan, India, and Nepal. Although *temperate japonica* and *tropical japonica* were identified as one varietal group, the results of this study are largely consistent with those of recent genomic studies.

These classifications of subgroups were confirmed by genetic studies utilizing a large number of samples and genome-wide genetic variation information, such as Single Sequence Repeat (SSR), Single Nucleotide Polymorphisms (SNPs) from randomly chosen gene fragments (Sequence-Tagged Sites (STSs)), and SNP panel assays (Garris et al. 2005; Caicedo et al. 2007; Zhao et al. 2010). Phylogenetic and STRUCTURE analyses indicated the existence of clear groupings for *temperate japonica*, *tropical japonica*, *aromatic* (including Basmati rice), *indica*, and *aus*.

temperate japonica, *tropical japonica* and *aromatic* were shown as the subgroups of ssp. *japonica*, and *temperate japonica* and *tropical japonica* showed closer genetic affinity and *aromatic* was placed on outside of them. *indica* and *aus* were shown as subgroups of the ssp. *indica*. In addition, a recent study employing whole-genome re-sequencing data for over 1000 samples confirmed that Rayada rice, a local group in Bangladesh, was clustered independently and placed between *japonica* and *indica* on the phylogenetic tree (Jing et al. 2023). Thus, the classification proposed by morphological features and isozyme analysis has been largely reconfirmed by studies based on genome-wide analyses using NGS technologies (Huang et al. 2012; Wang et al. 2018; Jing et al. 2023).

9.4 Geographic Origin of Cultivated Rice and Genomic Relationships of *japonica* and *indica*

The geographical origin and domestication history of Asian rice have been discussed for decades by a number of studies that have conducted phylogenetic analyses of cultivated and wild rice. There are two hypotheses on Asian rice domestication, single domestication and multiple domestication, and have been under discussion. The single domestication hypothesis assumes that *japonica* and *indica* originated from the same ancestral wild *O. rufipogon* population. In contrast, the multiple domestication hypothesis assumes that *japonica* and *indica* are domesticated separately from the different wild rice populations (Kovach et al. 2007; Sang and Ge 2007). Archaeological evidence indicates that early rice cultivation took place in two centers, one in the Middle or Lower Yangtze, China, and the other in the Middle Ganges valley, India (Fuller 2007). Together with insights from the current distribution range of varietal groups and genetic data, the former has been considered as the origin of *japonica*, and the latter as the origin of *indica* (Fig. 9.1). Focusing on geographical origin, this question is quite simple and is expected to be solved by studying the genetic relationships between *japonica*, *indica*, and *O. rufipogon* populations. Early studies conducted phylogeographic analysis using several neutral genetic loci for both cultivated and wild rice, demonstrated that *japonica* and *indica* diverged well and were closely related to different *O. rufipogon* populations. However, some studies on genes important for domestication traits found that *japonica* and *indica* shared the same alleles (Kovach et al. 2007; Sang and Ge 2007). Some studies utilizing genome-wide data have drawn incongruent conclusions supporting the single domestication or multiple domestication hypotheses as below. This is because the genomic relationships between *japonica*, *indica*, and *O. rufipogon* are complicated, reflecting the complex history of the domestication process resulting from human selection.

Early molecular phylogenetic studies based on genetic data examined genetically neutral regions not involved in the artificial selection process during domestication events, including chloroplast DNA as genetic markers, and found that *japonica* and *indica* are closely related to different *O. rufipogon* populations (Cheng

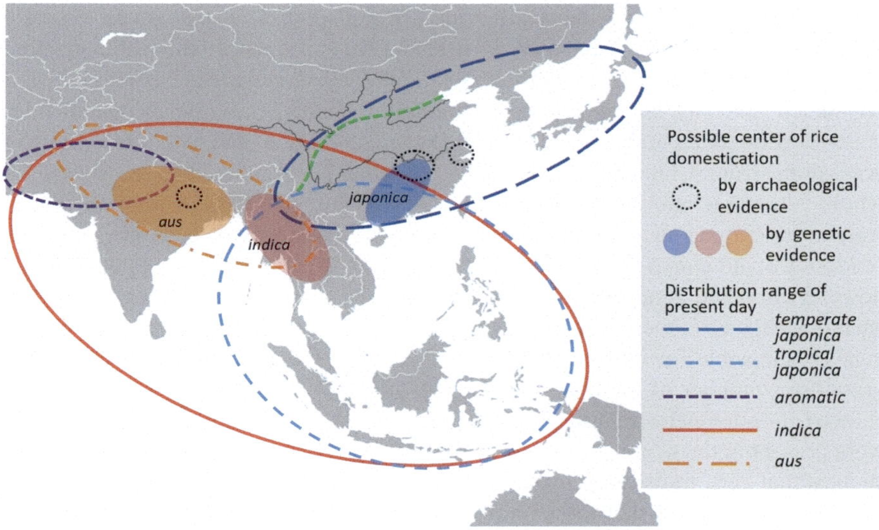

Fig. 9.1 Rough distribution of cultivated rice in Asia (Glaszmann 1987; Sweeny et al. 2007; Zhao et al. 2011), and the geographic origin of *japonica, indica* and *aus* inferred from archaeological evidence and genomics (Huang et al. 2012; Civáň et al. 2015; He et al. 2017; Jing et al. 2023)

et al. 2003; Londo et al. 2006; Rakshit et al. 2007). Londo et al. (2006) conducted a phylogeographic analysis of a large number of wild rice samples collected from a wide range of Asia, covering hypothetical domestication centers. For totally 203 cultivars (194–240 sequences) of *O. sativa* and 129 populations (94–677 sequences) of *O. rufipogon* that covered the entire range of geographic distributions of *O. rufipogon*, phylogenetic analyses of one chloroplast and two nuclear genes were conducted. These results suggest that rice domestication occurred at least twice in the different *O. rufipogon* populations. From the geographical distribution of alleles shared with *O. sativa* and *O. rufipogon*, the domestication origins of *japonica* and *indica* were suggested, *indica* was domesticated in South and Southeast Asia ranged from eastern India, Myanmar and Thailand, whereas *japonica* was domesticated in southern China. The divergence time of *japonica* and *indica* was also calculated based on DNA sequence differences (Ma and Bennetzen 2004; Vitte et al. 2004; Zhu and Ge 2005). The estimated divergence times were over 200,000 years ago, much before the start of the domestication process. These phylogeographic and divergence time estimation analyses using neutral genes, strongly suggest multiple domestications of Asian rice.

Subsequent studies utilizing population genetic analyses for genome-wide data showed a complicated signature of rice domestication history, indicating opposite conclusions supporting single domestication (Gao and Innan 2008; Molina et al. 2011) or multiple domestication (He et al. 2011; Yang et al. 2012). Molina et al. (2011) obtained sequences of 630 gene fragments on chromosomes 8, 10 and 12 for 20 *indica*, 16 *tropical japonica* and 20 *O. rufipogon* accessions. Demographic modeling of SNP data using a diffusion-based approach and a Bayesian evolutionary approach to phylogenetic analysis provided strong support for the complete single domestication of *japonica* and *indica*. He et al. (2011) sequenced the whole genome of 66 accessions (22 each from *japonica*, *indica*, and *O. rufipogon*) using NGS and conducted diversity and coalescence simulations. Yang et al. (2012) conducted a genome-wide comparative analysis of *japonica*, *indica*, and two *O. rufipogon* accessions from China and India with potential geographic origins. Based on whole-genome window analysis, these studies indicate that each gene in the *japonica* and *indica* genomes has a different history. The majority of genomic segments of *japonica* and *indica* showed closer affinity to different *O. rufipogon* populations; however, in some gene regions, *japonica* and *indica* were very closely related to each other than to other *O. rufipogon* populations. In particular, regions containing genes responsible for important traits for domestication, such as seed width, erect grass type, number of offshoots, seed color, amylose content, and seed dormancy, showed monophyletic relationships between *japonica* and *indica*. These regions were present in the genome as relatively large clusters. These observations strongly suggest that gene transfer occurred between *japonica* and *indica* during domestication. They concluded that the overall genome-wide pattern supports independent domestication of *japonica* and *indica*, while genomic segments showing low genetic diversity among *japonica* and *indica* cultivars containing important agronomic traits originated from one population and spread to all cultivars through introgression and human selection. This gene introgression was found to be primarily from *japonica* to *indica* and its extent was ~17% of the *indica* genome (Yang et al. 2012; Choi et al. 2017). In addition, ~15% of the *aus* genome was introgressed from *japonica* (Choi et al. 2017). The proposed complex genomic relationships of the *japonica*, *indica*, and *O. rufipogon* populations were confirmed by large-scale genome re-sequencing studies employing over 1000 samples of cultivated and wild rice (Xu et al. 2011; Huang et al. 2012).

Huang et al. (2012) tried to pinpoint the domestication origin of *japonica* group in China employing more than a thousand samples, 1083 cultivated rice and 446 wild ancestor of cultivated rice *O. rufipogon*. The detailed location of *japonica* origin was estimated based on the distribution of the *O. rufipogon* population, which has the closest genetic affinity to cultivated rice in domestication-related gene regions where selective sweeps were identified. They concluded that *japonica* was first domesticated from the *O. rufipogon* population in the middle area of the Pearl River in southern China. However, from the context of archaeological findings, this area has not been assumed as a candidate location for the domestication of *japonica* started (Gross and Zhao 2014). It is possible that gene flow from cultivated to wild rice affected this inconsistency (Wang et al. 2017). In addition, *indica* was estimated to have developed by crossing of *japonica* and local wild rice after *japonica* spread into South and Southeast Asia (Huang et al. 2012). Recently, Jing et al. (2023) analyzed 1578 resequenced genomes, of which 459 were newly obtained genomes with a relatively high depth of 11.7×. They enriched wild rice collections of *O. rufipogon* and *O. nivara* (also known as the annual ecotype of *O. rufipogon*) to investigate the ancestry of the domestication genes. By investigating the topologies of phylogenetic trees, they found that ~80% of selected genes have ancestry in wild rice in China, whereas ~20% of them originated from wild rice in South and Southeast Asia. They concluded that the domestication of *japonica* and *indica* occurred independently and was completed by continuous selection and bi-directional genetic exchanges.

The geographical origins of other varietal groups have also been investigated. Origin of *aus* was investigated by utilizing variation data of Huang et al. (2012) and suggested that they are independently domesticated in central India or Bangladesh (Civáň et al. 2015). Furthermore, aromatic rice was suggested to arose from hybridization between *japonica* that had spread from East Asia and *aus*. The genetic contribution of the local wild rice population in the Indian sub-continent has also been demonstrated (Civáň et al. 2015, 2019). Most of the aromatics had cytoplasm, and 29–47% of the nuclear genome was derived from local rice in India. The admixture event between them was estimated to be 4000–2400 years BP.

As shown above, even though a vast amount of genomic data has been analyzed, two hypotheses on the rice domestication process are under discussion. However, the focus of the discussion on single and multiple domestication hypotheses has evolved into more complicated genomic stories. The genomic backgrounds of *japonica* and *indica*, and their mutual influence on domestication genes is debated as introgression and replacement models (Fig. 9.2) (Sang and Ge 2013; Callaway 2014). In the former model, *japonica* and *indica* were independently domesticated in China and Southeast Asia, respectively, and genes related to domestication were transferred from *japonica* to *indica* by hybridization. In the latter model, *japonica* was first domesticated and then brought to Southeast Asia, where it was crossed with local wild rice or primitive *indica*, and then gradually replaced the genomic background with domestication-related gene regions. Both scenarios are possible, but the former seems to be simpler (Wang et al. 2017).

A different approach focused on the repertoire of domestication-associated genes, *qSW5*, *Rc*, *qSH1*, *sh4*, *waxy*, and *Rd*, to determine the origin of cultivated rice (Shomura et al. 2008; Konishi et al. 2008). The distribution of 'heritage landrace' with a combination of wild-type alleles without domestication mutations was found to be in South-East Asia, the Philippines, Indonesia and parts of Indochina, suggesting the origin of *japonica* rice in these regions. However, this conclusion is

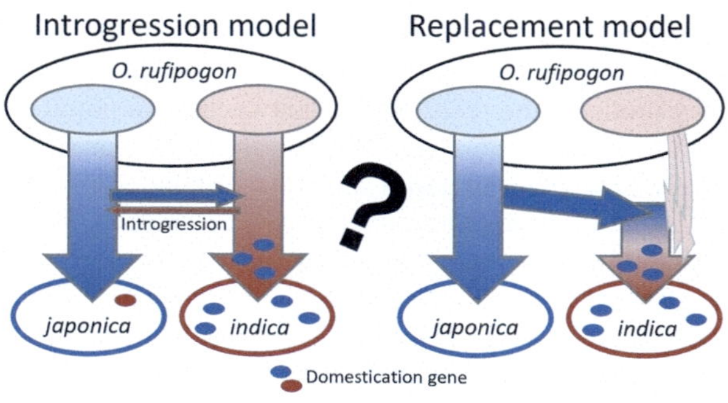

Fig. 9.2 Two models explaining the sharing of genomic segments especially domestication genes between *japonica* and *indica*

unsuitable for archeological data from paddy ruins and the history of human migration in Southeast Asia (Fuller and Sato 2008; McColl et al. 2018; Lipson et al. 2018). When studying the origin of Asian rice, we should consider the following possibilities: (1) differences in the distribution of wild ancestors between the present and the past; (2) cultivated rice varieties are easily transmitted through trade and human immigration; (3) drastic changes in the varieties used as a result of climate change and human activity; and (4) mutual genetic exchange between wild and cultivated rice in areas where wild rice is native. Therefore, it is important not only to discuss the genetic aspects, but also to take a composite perspective in conjunction with advances in fields such as archaeology, linguistics, and folklore to clarify the origin and evolutionary process of rice domestication.

9.5 Dispersal History of Cultivated Rice Populations

The largest genomics project in rice research to date, the 3000 rice genomes project sequenced a worldwide cultivated rice collection from 89 countries at an average depth of 14× (3,000 rice genomes project 2014). A study that revealed comprehensive genomic diversity using resequenced data from these 3010 varieties found nine *indica* and *japonica* subpopulations, which roughly corresponded to their geographical distribution (Wang et al. 2018). Four *indica* subpopulations: XI-1A from East Asia, XI-1B of the modern variety, XI-2 from South Asia, and XI-3 from Southeast Asia, and three *japonica* subpopulations: *temperate japonica* from East Asia, subtropical, and *tropical japonica* from Southeast Asia. The other two groups were South Asian circum-*Aus* (cA) and circum-*Basmati* (cB); the former showed close affinity to *indica*, and the latter was close to *japonica*.

Several recent studies have utilized genomic data of more than 1000 varieties from the 3000 rice genomes project to infer demographic history through detailed population genomic analysis (Gutaker et al. 2020; Alam et al. 2021; Liu et al. 2023). Gutaker et al. (2020) conducted population genomic analysis using genomic data of 1400 varieties and combined with geographic, environmental archaeobotanical and paleoclimate data to estimate the origins of regional populations and process of geographical spreading of *japonica* and *indica* (Fig. 9.3). This is the first comprehensive study to demonstrate the detailed relationships between geographical populations of Asian rice. Based on the admixture-graph framework, they modeled the relationships between local populations and estimated their dispersal history. Most importantly, this study indicated that *temperate japonica* emerged from an ancestral *japonica* population that had basic tropical components after the 4.2 k global cooling event. The split of *temperate* and *tropical japonica* was estimated to start at approximately 4100 years BP by cross-coalescence analysis. The 4.2 k cooling event might pressure the *japonica* population and result in *japonica* rice being adapted to a temperate climate. A reduction in population size (*N*e) was observed at 3500–3000 years BP in *temperate japonica*. Together with the increase in archaeological sites with rice remains in Northeast Asia, this can be

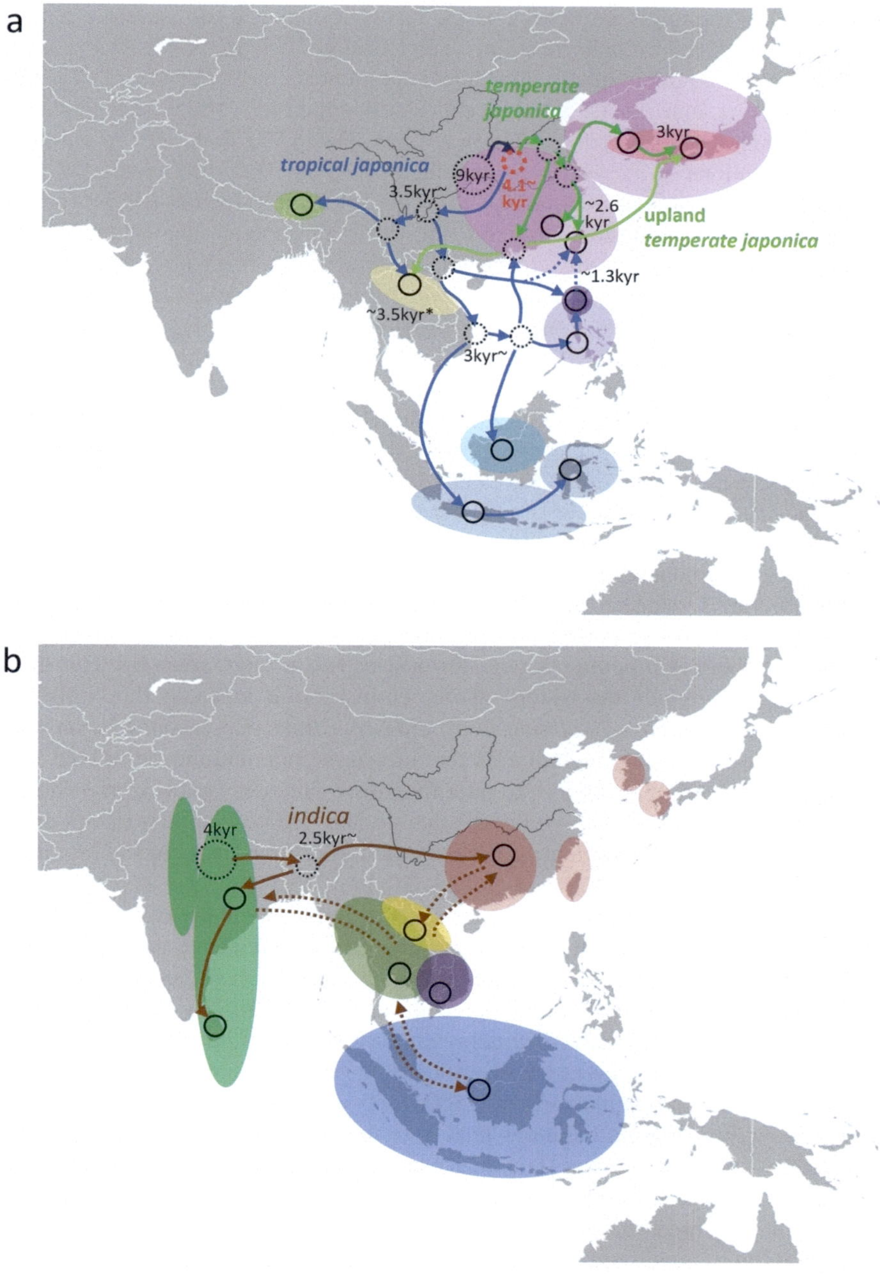

Fig. 9.3 Dispersal map of (**a**) *japonica* and (**b**) *indica* populations. The solid circles indicate center of geographical populations highlighted with colors. Dashed circles indicate hypothetical ancestral populations during dispersal process. Their precise geographical locations are uncertain. Arrows indicate proposed hypothetical route. (Adopted from Gutaker et al. (2020) and Alam et al. (2021))

interpreted as a founder bottleneck as the result of dispersal of *temperate japonica* to Northeast Asia, finally reaching Japan around 3000 years BP. At the same time, the *japonica* population with tropical components has spread from China to Southeast Asia and South Asia. Southeast populations (Malay Archipelago) diverged from the mainland population around 3000–1500 years BP. In addition, island populations of the Malay Archipelago diverged 3000–1000 years BP. The spread to Southeast Asia was estimated to be older, ~3500 years BP, by further research based on additional datasets and different analysis algorithms (Alam et al. 2021). It is noteworthy that tropical components from the island Malay Archipelago were mixed with temperate components, and reached Japan as upland *temperate japonica* (note: sometimes recognized as *tropical japonica*) through Taiwan and the Ryukyu Islands (Gutaker et al. 2020). The genomic components of the Taiwanese rice population were shown to be complex and two distinct populations were identified (Alam et al. 2021). Both populations had their half of the genomic component from the Northeast Asian *temperate japonica* population, including varieties from China, Korea and Japan. And one population had the contribution of *tropical japonica* from the northern Philippines and clustered well with the upland *temperate japonica* of Northeast Asia. The other population also had tropical components from mainland Southeast Asia. The *temperate japonica* components of Taiwanese rice diverged from the Northeast Asian population at approximately 2600 years BP, and *tropical japonica* components were derived from the northern Philippines as gene flow started before 1300 years BP. This is an interesting result given the history of rice cultivation in Japan. If the Northeast Asian upland *temperate japonica* consisting of China, Korea, and Japan was derived from the same population as the Taiwanese population, the upland rice might have arrived in Japan a little later than the *temperate japonica* population. This is inconsistent with the hypothesis that upland rice was the first rice population cultivated in Japan during the Yayoi period. Additional analyses using more landrace samples from this area are required.

The relationship and dispersal history of the *indica* populations seemed to be more complicated. Although *indica* varieties were clustered into six populations, they showed a weak population structure, which might have been caused by continuous gene flow between populations (Gutaker et al. 2020). The graph topologies were highly diversified and reconstruction of the dispersal route of the *indica* population was difficult. However, the separation of the Chinese and Indian populations could be confidently estimated to be 2500–1100 years BP. The possible dispersal routes of *indica* rice from India to China were the Silk Road and/or direct passage to southwest China across the Hengduan Mountains located in the southeast of the Tibetan Plateau (Gutaker et al. 2020). The dispersal route to Southeast Asia, from either India or China, was not clear. Archaeological evidence suggests that *indica* rice arrived in Thailand at 1800 years BP (Castillo et al. 2018).

One recent study attempted to shed light on the relationships of local rice populations in a narrower geographical area based on population genomics analysis (Liu et al. 2023). Through history of the rice variety, they tried to determine the origin of ethnic minorities in China. A comparative genome analysis of Kam Sweet Rice, a distinctive rice variety used uniquely by the Dong people in China, with other landraces from the south of the Yangtze River Basin in China was conducted. The results provide evidence for the southeast coast origin of the Dong people at least 1000 years BP. In addition, they identified that genes related to agronomic traits *GS3*, *Hd1*, and *DPS1*, and those related to cold tolerance *LTG1* and *MYBS3* were domestication targets for Kam Sweet Rice. These genes were selected for crop improvement and adaptation to the geographical environment changes where the Dong people migrated. Conducting this kind of high-resolution study on local rice populations, including further data accumulation and integration of existing data, will reveal a more detailed history of rice dispersal in the future.

9.6 Genomic Diversity of Cultivated Rice in Japan

Fifty varieties from the National Agriculture and Food Research Organization (NARO) Japanese Rice Collection (JRC) were sequenced at a relatively thick average depth of ×29 (Tanaka et al. 2021). This rice collection was developed from NARO's collection of 2000 varieties of Japanese rice landraces, with confirmed provenance information and a broad coverage of the genetic diversity of Japanese rice (Kojima et al. 2005; Ebana et al. 2008). Molecular phylogenetic analysis of whole genome data in conjunction with similar high depth genomic data from NARO's World Rice Collection (WRC) (Tanaka et al. 2020b) identified three subgroups, *temperate japonica*, *tropical japonica* and *indica*, in the Japanese native rice landrace collection varieties. Genomics data of the JRC, WRC, and 20 representative varieties from major groups of the 3000 rice genomes project were reanalyzed. A molecular phylogenetic tree was constructed using the whole genome SNPs (Fig. 9.4). *temperate japonica* and *tropical japonica*, noted as upland rice, clearly showed distinct monophyletic clusters. There were four *indica* varieties in the JRC, and they were clustered with *indica* varieties belonging to XI-1A, which is the major group of East Asia, mainly China (Wang et al. 2018). This result was consistent with the written in old records that some *indica* varieties were thought to have been introduced from China at fourteenth to fifteenth century (Itani and Ogawa 2004). Actually, some of these varieties have names like "Touboshi" and "Karahoushi", the "Tou" and "Kara" means T'ang in Chinese character *indica*ting China or T'ang dynasty (seventh to tenth centuries). Other varieties clustered in the *indica* group were named "Akamai" meaning red rice. "Akamai" is a so-called ancient rice variety, with several cultivation traits that are primitive and close to the wild type, recalling varieties used by ancient peoples due to their characteristics (Arashi 1974). The first coherent description of this lineage dates back to 1671 and has been recognized for a long time. It is assumed that the *japonica* type of "Akamai" came to Japan earlier and was widely cultivated in Japan in the seventh to eighth centuries (Asuka period) from the records of wooden tablets excavated from archaeological sites in Heijou-kyou, the ancient capital city of Nara (Itani and Ogawa 2004). There were two Akamai varieties in the *japonica* cluster: one was in the *temperate*

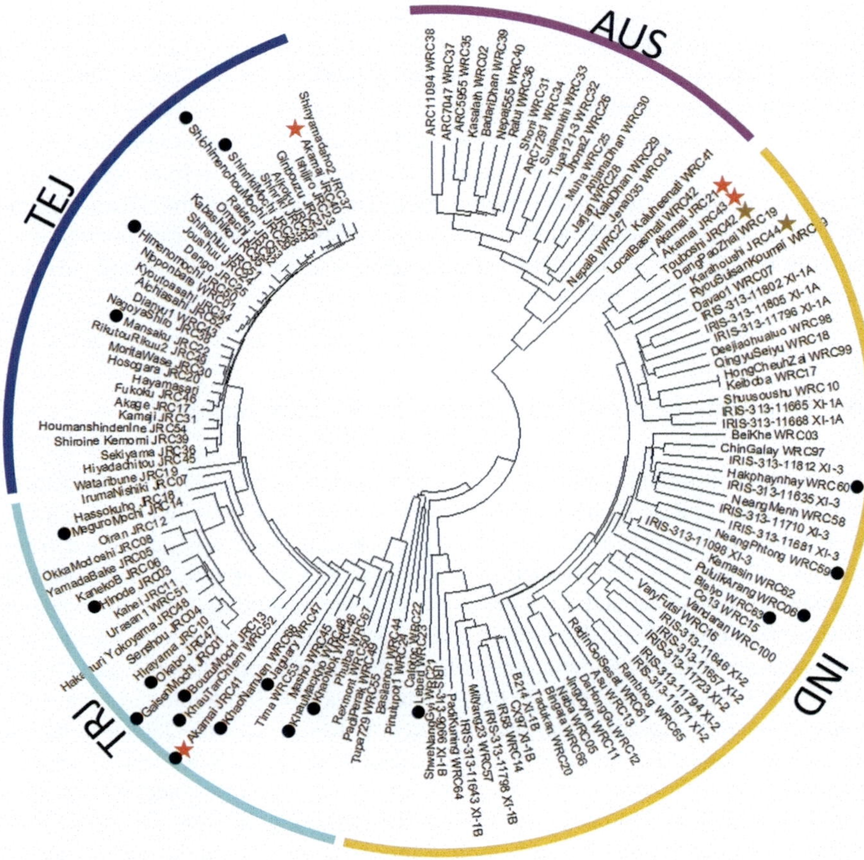

Fig. 9.4 Phylogenetic tree of whole genome SNPs data for WRC, JRC (Japanese rice collection) and representative accessions from Wang et al. (2018). Red stars indicate "Akamai" accessions in JRC and brown stars indicate *indica* accessions in JRC. Filled black circles indicate glutinous rice in JRC and WRC

japonica and it branched at the very terminal branch, and the other, JRC41, diverged earlier from the *tropical japonica* cluster, suggesting that it was a very old variety. This variety may be the ancient "Akamai" variety that dates back to the seventh century. Therefore, a detailed analysis of the domestication gene region was performed (Table 9.2, Fig. 9.5). By investigating the phylogenetic tree for 13 genes, it is found that JRC41 had older allele branched earlier in *japonica* for 7 genes. For these four genes, typical *indica*-type alleles were observed in JRC41. These results suggested that JRC41 "Akamai" variety has genomic component diverged earlier in *japonica* variety but experienced hybridization with *indica*.

Admixture analysis of genomic data revealed two subpopulations of *temperate japonica* in Japan (Fig. 9.6a). Plots of the location where each variety was collected showed a trend towards differential distribution between the north and south of Japan, namely TEJ_N and TEJ_S (Fig. 9.6b). Alternatively, it can be interpreted that

Table 9.2 Genotype of domestication genes for JRC41 "Akamai"

Gene	Phenotype	Genotype of JRC41
SH4	Seed shattering	JP-old
qSH1	Seed shattering	IND
SDR4	Seed dormancy	JP
PROG1	Erect growth	JP-old
qSW5	Grain width	JP-old
AN1	Presence/absence of awn	JP-old
LG1, OsLG1	Inflorescence formation	IND
wx	Amylose content	JP-old
RC	Seed color	JP-old
Rd1/Rd2	Seed color	IND
BH4	Seed color	JP
BADH-2	Stress tolerance	JP-old
HAN1	Stress tolerance	IND

For Genotype, JP indicate typical *japonica* type, JP-old indicates allele branched earlier and IND indicates typical *indica* type as shown in Fig. 9.5

they are distributed in the center and periphery of the Japanese archaepelago. Although further analysis is required owing to the small sample size, it would be interesting if there was a structure within the Japanese rice varieties. When looking at whether there were differences in any phenotypic characteristics between these populations, significant differences were found in the awn phenotype (Fig. 9.6c), which is involved in seed dispersal and protection against seed predation and is important for wild rice (Grundbacher 1963). In the domestication process of rice, the awnless phenotype is thought to be advantageous and is selected to facilitate planting, harvesting, and storage of seeds (Tatsumi and Kawano 1972). Genome-wide F_{ST} values between the two populations showed several diversified regions with high F_{ST} tracks, including one particularly high peak on chromosome 8, where a gene encodes the Regulator of Awn Elongation 2 (RAE2) protein (Fig. 9.7). RAE2 was responsible for awn less, and a selective sweep was also detected at this locus, indicating that the awnless-type phenotype was a target for selection during human cultivation (Bessho-Uehara et al. 2016). The nearby regions of this locus showed a very high peak associated with awn length in a GWAS study of *temperate japonica* varieties (Yano et al. 2016). In the present study, when we examined the *RAE2* gene region of the two subpopulations with the genome browser TASUKE+ (Kumagai et al. 2019), polymorphisms that alter the amino acid sequence and functional alleles that develop awn were highly prevalent in the TE_N population, 10 out of 12 varieties (83%). A broader look at contemporary variety data shows that loss of function alleles for the *RAE2* gene are also very frequent in both *japonica* and *indica* groups outside Japan (RAP-DB TASUKE+ for 685 varieties: https://agrigenome.dna.affrc.go.jp/tasuke/ricegenomes/). As the presence of awn is one of the characteristics of old rice varieties, the Japanese TEJ_N population may have retained the traits of the old lineages. For other high-F_{ST} regions, there was no overlap with other agronomic trait loci (for example. Heading date, plant height, leaf blade width, panicle length, spikelet number per panicle, leaf blade width, and

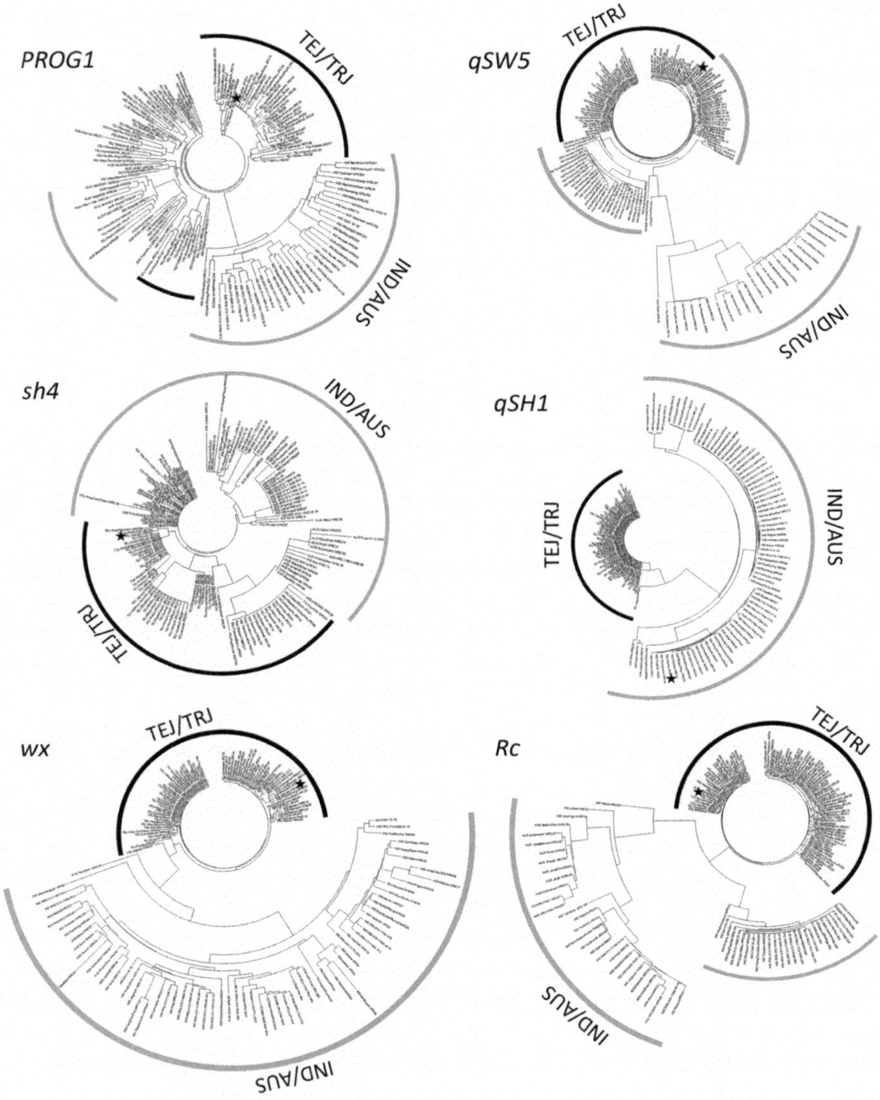

Fig. 9.5 Phylogenetic tree of domestication genes (*PROG1*, *qSW5*, *sh4*, *qSH1*, *wx* and *Rc*) for WRC, JRC and representative accessions from Wang et al. (2018). Black stars indicate the *tropical japonica* type "Akamai" accession. *temperate japonica* and *tropical japonica* alleles (TEJ/TRJ) are indicated by black arcs and *indica* and *aus* alleles (IND/AUS) are indicated by grey arcs

panicle length) reported in previous GWAS studies (Yano et al. 2016; Tanaka et al. 2021). As data on various traits will be obtained for the NARO core collection, it is expected that some traits related to the genomic regions that diverged in the two subpopulations will be found in the future. It is an interesting finding in the history of rice introduction to Japan that, in addition to the two previously discussed

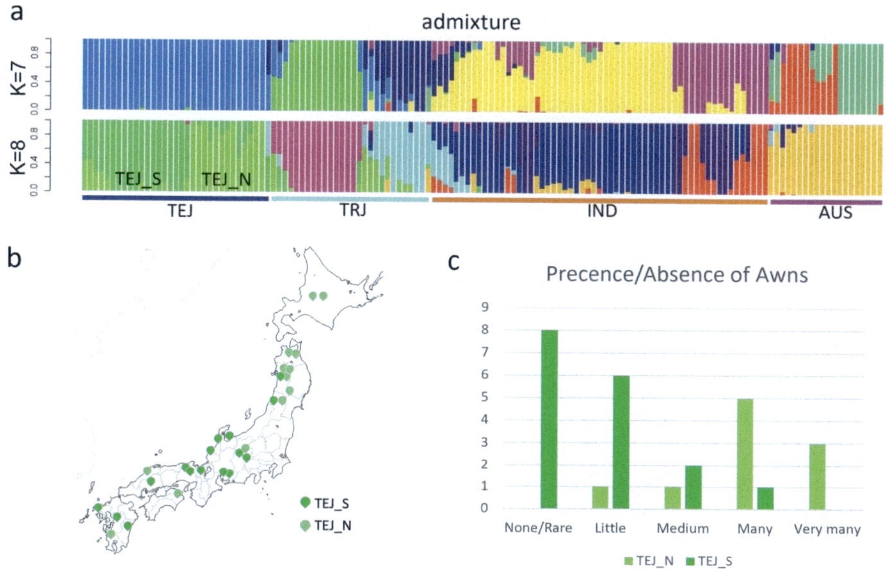

Fig. 9.6 (**a**) Population structure for WRC, JRC and representative accessions from Wang et al. (2018). (**b**) Distribution of subpopulation (TEJ_N and TEJ_S) recognized within *temperate japonica* in Japan. (**c**) Awn phenotypic value for two subpopulations

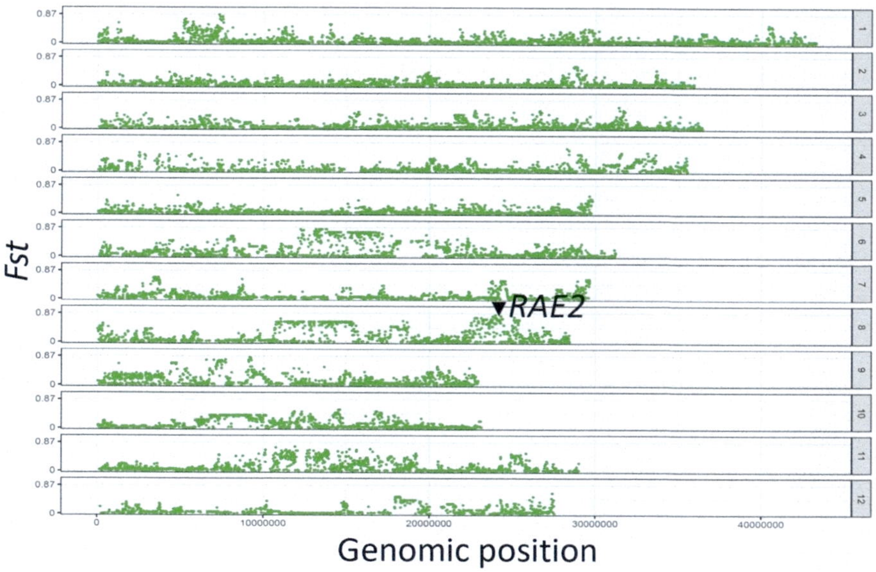

Fig. 9.7 Genome wide distribution of *Fst* between two subpopulations in Japanese *temperate japonica* (TEJ_N and TEJ_S). Black triangle indicates *RAE2* gene responsible for awn phenotype

varietal groups of *tropical japonica* and *temperate japonica*, it has been suggested that *temperate japonica* in Japan may have originated from more than one source.

9.7 Perspectives

Although the acquisition of large-scale genomic data and advances in population genome analysis have revealed many findings about the origins and population histories of cultivated rice in Asia, we are only just getting started as there are still many areas where conclusions are not yet certain. An overview of the diffusion of *japonica* and *indica* populations across Asia is presented and interpreted by combining archaeological and climatic data is being attempted. A recent study aimed to determine the relationships between narrow regional populations in China. The population history of rice in Japan is also expected to become clearer in comparison with abundant data from China and Korea. In addition, even with the availability of large amounts of genomic data, it may still be difficult to completely unravel the historical story from a modern sample alone. By analyzing rice remains and carbonized rice excavated from archaeological sites, it is possible to directly clarify the varieties used in ancient times and their characteristics. Ancient DNA analysis is a powerful tool for gaining new insights. For rice, some studies on DNA analysis of carbonized rice using chloroplast DNA markers have been reported, and genetic diversity in past rice varieties was disclosed (Castillo et al. 2016; Kumagai et al. 2016; Tanaka et al. 2016, 2020a). Ancient genomics analysis has been successful in other crop species, such as maize, wheat, barley, and common bean (Mascher et al. 2016; Ramos-Madrigal et al. 2016; Vallebueno-Estrada et al. 2016; Scott et al. 2019; Trucchi et al. 2021). Currently, the success of genome decoding depends on the proportion of endogenous DNA present in ancient remains. In successful research cases, archaeological samples appear to be very well conserved under dry conditions, and they seem to be not carbonized. So far, it has been difficult to conduct genomic analysis for carbonized rice because of the very low endogenous DNA ratio. However, it is hoped that innovation in experimental techniques and the search for samples in good condition can lead to success of rice ancient genomics, as has begun to succeed in the difficult analysis of ancient East Asian human genomics in the Yaponesia genome project.

Acknowledgements We thank Dr. Takeshi Itoh and Dr. Yoshihiro Kawahara for fruit full discussions. This work was supported by MEXT KAKENHI (grants 21K00996 and 20H05820 for MK and 18H05508 for HS).

Competing InterestsThe authors have no relevant financial or non-financial interests to disclose.

References

3,000 rice genomes project (2014) The 3,000 rice genomes project. Gigascience 3:7

Aggarwal RK, Brar DS, Khush GS (1997) Two new genomes in the *Oryza* complex identified on the basis of molecular divergence analysis using total genomic DNA hybridization. Mol Gen Genet 254(1):1–12

Aggarwal RK, Brar DS, Nandi S, Huang N, Khush GS (1999) Phylogenetic relationships among *Oryza* species revealed by AFLP markers. Theor Appl Genet 98:1320–1328

Alam O, Gutaker RM, Wu CC, Hicks KA, Bocinsky K, Castillo CC, Acabado S, Fuller D, d'Alpoim Guedes JA, Hsing YI, Purugganan MD (2021) Genome analysis traces regional dispersal of rice in Taiwan and Southeast Asia. Mol Biol Evol 38(11):4832–4846

Arashi K (1974) The Japanese red rice study (In Japanese). Yuzankaku Publishers, Tokyo

Bessho-Uehara K, Wang DR, Furuta T, Minami A, Nagai K, Gamuyao R, Asano K, Angeles-Shim RB, Shimizu Y, Ayano M, Komeda N, Doi K, Miura K, Toda Y, Kinoshita T, Okuda S, Higashiyama T, Nomoto M, Tada Y, Shinohara H, Matsubayashi Y, Greenberg A, Wu J, Yasui H, Yoshimura A, Mori H, McCouch SR, Ashikari M (2016) Loss of function at *RAE2*, a previously unidentified EPFL, is required for awnlessness in cultivated Asian rice. Proc Natl Acad Sci U S A 113(32):8969-74

Caicedo AL, Williamson SH, Hernandez RD, Boyko A, Fledel-Alon A, York TL, Polato NR, Olsen KM, Nielsen R, McCouch SR, Bustamante CD, Purugganan MD (2007) Genome-wide patterns of nucleotide polymorphism in domesticated rice. PLoS Genet 3(9):1745–1756

Callaway E (2014) Domestication: the birth of rice. Nature 514(7524):S58–S59

Castillo CC, Tanaka K, Sato YI, Ishikawa R, Bellina B, Higham C, Chang N, Mohanty R, Kajale M, Fuller DQ (2016) Archaeogenetic study of prehistoric rice remains from Thailand and India: evidence of early *japonica* in south and Southeast Asia. Archaeol Anthropol Sci 8:523–543

Castillo CC, Higham CFW, Miller K, Chang N, Douka K, Higham TFG, Fuller DQ (2018) Social responses to climate change in Iron Age north-east Thailand: new archaeobotanical evidence. Antiquity 92:1274–1291

Cheng C, Motohashi R, Tsuchimoto S, Fukuta Y, Ohtsubo H, Ohtsubo E (2003) Polyphyletic origin of cultivated rice: based on the interspersion pattern of SINEs. Mol Biol Evol 20(1):67–75

Choi JY, Platts AE, Fuller DQ, Hsing YI, Wing RA, Purugganan MD (2017) The rice paradox: multiple origins but single domestication in Asian rice. Mol Biol Evol 34(4):969–979

Civáň P, Craig H, Cox CJ, Brown TA (2015) Three geographically separate domestications of Asian rice. Nat Plants 1:15164

Civáň P, Ali S, Batista-Navarro R, Drosou K, Ihejieto C, Chakraborty D, Ray A, Gladieux P, Brown TA (2019) Origin of the aromatic group of cultivated rice (*Oryza sativa* L.) traced to the Indian subcontinent. Genome Biol Evol 11(3):832–843

Du H, Yu Y, Ma Y et al (2017) Sequencing and de novo assembly of a near complete *indica* rice genome. Nat Commun 8:15324

Ebana K, Kojima Y, Fukuoka S, Nagamine T, Kawase M (2008) Development of mini core collection of Japanese rice landrace. Breed Sci 58:281–291

Fuller DQ (2007) Contrasting patterns in crop domestication and domestication rates: recent archaeobotanical insights from the Old World. Ann Bot 100:903–924

Fuller DQ, Sato YI (2008) *Japonica* rice carried to, not from, Southeast Asia. Nat Genet 40:1264–1265

Gao LZ, Innan H (2008) Nonindependent domestication of the two rice subspecies, *Oryza sativa* ssp. *indica* and ssp. *japonica*, demonstrated by multilocus microsatellites. Genetics 179(2):965–976

Garris AJ, Tai TH, Coburn J, Kresovich S, McCouch S (2005) Genetic structure and diversity in *Oryza sativa* L. Genetics 169(3):1631–1638

Ge S, Sang T, Lu BR, Hong DY (1999) Phylogeny of rice genomes with emphasis on origins of allotetraploid species. Proc Natl Acad Sci U S A 96(25):14400–14405

Glaszmann JC (1987) Isozymes and classification of Asian rice varieties. Theor Appl Genet 74(1):21–30

Gross BL, Zhao Z (2014) Archaeological and genetic insights into the origins of domesticated rice. Proc Natl Acad Sci U S A 111(17):6190–6197

Grundbacher FJ (1963) The physiological function of the cereal awn. Bot Rev 29:366–381

Gutaker RM, Groen SC, Bellis ES, Choi JY, Pires IS, Bocinsky RK, Slayton ER, Wilkins O, Castillo CC, Negrão S, Oliveira MM, Fuller DQ, Guedes JAD, Lasky JR, Purugganan MD (2020) Genomic history and ecology of the geographic spread of rice. Nat Plants 6(5):492–502

He Z, Zhai W, Wen H, Tang T, Wang Y, Lu X, Greenberg AJ, Hudson RR, Wu CI, Shi S (2011) Two evolutionary histories in the genome of rice: the roles of domestication genes. PLoS Genet 7(6):e1002100

He K, Lu H, Zhang J, Wang C, Huan X (2017) Prehistoric evolution of the dualistic structure mixed rice and millet farming in China. The Holocene 27:12

Huang X, Kurata N, Wei X, Wang ZX, Wang A, Zhao Q, Zhao Y, Liu K, Lu H, Li W, Guo Y, Lu Y, Zhou C, Fan D, Weng Q, Zhu C, Huang T, Zhang L, Wang Y, Feng L, Furuumi H, Kubo T, Miyabayashi T, Yuan X, Xu Q, Dong G, Zhan Q, Li C, Fujiyama A, Toyoda A, Lu T, Feng Q, Qian Q, Li J, Han B (2012) A map of rice genome variation reveals the origin of cultivated rice. Nature 490(7421):497–501

International Rice Genome Sequencing Project (2005) The map-based sequence of the rice genome. Nature 436:793–800

Itani T, Ogawa M (2004) History and recent research situation of red rice cultivation in Japan. Bull Crop Sci Soc Jpn 73(2):137–147. (in Japanese)

Jing CY, Zhang FM, Wang XH, Wang MX, Zhou L, Cai Z, Han JD, Geng MF, Yu WH, Jiao ZH, Huang L, Liu R, Zheng XM, Meng QL, Ren NN, Zhang HX, Du YS, Wang X, Qiang CG, Zou XH, Gaut BS, Ge S (2023) Multiple domestications of Asian rice. Nat Plants 9(8):1221–1235

Joshi SP, Gupta VS, Aggarwal RK, Ranjeker PK, Brar DS (2000) Genetic diversity and phylogenetic relationship as revealed by inter simple sequence repeat (ISSR) polymorphism in the genus *Oryza*. Theor Appl Genet 100:1311–1320

Kato S, Kosaka H, Hara S (1928) On the affinity of rice varieties as shown by the fertility of rice plants. Centr Agric Inst Kyushu Imp Univ 2:241–276

Kawahara Y, de la Bastide M, Hamilton JP, Kanamori H, McCombie WR, Ouyang S, Schwartz DC, Tanaka T, Wu J, Zhou S, Childs KL, Davidson RM, Lin H, Quesada-Ocampo L, Vaillancourt B, Sakai H, Lee SS, Kim J, Numa H, Itoh T, Buell CR, Matsumoto T (2013) Improvement of the *Oryza sativa* Nipponbare reference genome using next generation sequence and optical map data. Rice (N Y) 6(1):4

Khush GS (1997) Origin, dispersal, cultivation and variation of rice. Plant Mol Biol 35(1–2):25–34

Kojima Y, Ebana K, Fukuoka S, Nagamine T, Kawase M (2005) Development of an RFLP-based rice diversity research set of germplasm. Breed Sci 55:431–440

Konishi S, Ebana K, Izawa T (2008) Inference of the *japonica* rice domestication process from the distribution of six functional nucleotide polymorphisms of domestication-related genes in various landraces and modern cultivars. Plant Cell Physiol 49(9):1283–1293

Kovach MJ, Sweeney MT, McCouch SR (2007) New insights into the history of rice domestication. Trends Genet 23(11):578–587

Kumagai M, Wang L, Ueda S (2010) Genetic diversity and evolutionary relationships in genus *Oryza* revealed by using highly variable regions of chloroplast DNA. Gene 462(1–2):44–51

Kumagai M, Kanehara M, Shoda S, Fujita S, Onuki S, Ueda S, Wang L (2016) Rice varieties in archaic East Asia: reduction of its diversity from past to present times. Mol Biol Evol 33(10):2496–2505

Kumagai M, Nishikawa D, Kawahara Y, Wakimoto H, Itoh R, Tabei N, Tanaka T, Itoh T (2019) TASUKE+: a web-based platform for exploring GWAS results and large-scale resequencing data. DNA Res 26(6):445–452

Li HW (1964) Studies on genetic and cytogenetic evidence for species relationships in the Republic of China. In: Tsunoda S, Takahashi N (eds) Rice genetics and cytogenetics. Elsevier, Amsterdam, pp 118–131

Lipson M, Cheronet O, Mallick S, Rohland N, Oxenham M, Pietrusewsky M, Pryce TO, Willis A, Matsumura H, Buckley H, Domett K, Nguyen GH, Trinh HH, Kyaw AA, Win TT, Pradier B, Broomandkhoshbacht N, Candilio F, Changmai P, Fernandes D, Ferry M, Gamarra B, Harney E, Kampuansai J, Kutanan W, Michel M, Novak M, Oppenheimer J, Sirak K, Stewardson K, Zhang Z, Flegontov P, Pinhasi R, Reich D (2018) Ancient genomes document multiple waves of migration in Southeast Asian prehistory. Science 361(6397):92–95

Liu C, Wang T, Chen H, Ma X, Jiao C, Cui D, Han B, Li X, Jiao A, Ruan R, Xue D, Wang Y, Han L (2023) Genomic footprints of Kam Sweet Rice domestication indicate possible migration routes of the Dong people in China and provide resources for future rice breeding. Mol Plant 16(2):415–431

Londo JP, Chiang YC, Hung KH, Chiang TY, Schaal BA (2006) Phylogeography of Asian wild rice, *Oryza* rufipogon, reveals multiple independent domestications of cultivated rice, *Oryza sativa*. Proc Natl Acad Sci U S A 103(25):9578–9583

Ma J, Bennetzen JL (2004) Rapid recent growth and divergence of rice nuclear genomes. Proc Natl Acad Sci U S A 101(34):12404–12410

Mascher M, Schuenemann VJ, Davidovich U, Marom N, Himmelbach A, Hübner S, Korol A, David M, Reiter E, Riehl S, Schreiber M, Vohr SH, Green RE, Dawson IK, Russell J, Kilian B, Muehlbauer GJ, Waugh R, Fahima T, Krause J, Weiss E, Stein N (2016) Genomic analysis of 6,000-year-old cultivated grain illuminates the domestication history of barley. Nat Genet 48(9):1089–1093

Matsumoto T, Wu J, Itoh T, Numa H, Antonio B, Sasaki T (2016) The Nipponbare genome and the next-generation of rice genomics research in Japan. Rice (N Y) 9(1):33

Matsuo T (1952) Genecological studies on cultivated rice (in Japanese). Bull Nat Inst Agric Sci Jpn D 3:1–111

McColl H, Racimo F, Vinner L, Demeter F, Gakuhari T, Moreno-Mayar JV, van Driem G, Gram Wilken U, Seguin-Orlando A, de la Fuente Castro C, Wasef S, Shoocongdej R, Souksavatdy V, Sayavongkhamdy T, Saidin MM, Allentoft ME, Sato T, Malaspinas AS, Aghakhanian FA, Korneliussen T, Prohaska A, Margaryan A, de Barros Damgaard P, Kaewsutthi S, Lertrit P, Nguyen TMH, Hung HC, Minh Tran T, Nghia Truong H, Nguyen GH, Shahidan S, Wiradnyana K, Matsumae H, Shigehara N, Yoneda M, Ishida H, Masuyama T, Yamada Y, Tajima A, Shibata H, Toyoda A, Hanihara T, Nakagome S, Deviese T, Bacon AM, Duringer P, Ponche JL, Shackelford L, Patole-Edoumba E, Nguyen AT, Bellina-Pryce B, Galipaud JC, Kinaston R, Buckley H, Pottier C, Rasmussen S, Higham T, Foley RA, Lahr MM, Orlando L, Sikora M, Phipps ME, Oota H, Higham C, Lambert DM, Willerslev E (2018) The prehistoric peopling of Southeast Asia. Science 361(6397):88–92

Molina J, Sikora M, Garud N, Flowers JM, Rubinstein S, Reynolds A, Huang P, Jackson S, Schaal BA, Bustamante CD, Boyko AR, Purugganan MD (2011) Molecular evidence for a single evolutionary origin of domesticated rice. Proc Natl Acad Sci U S A 108(20):8351–8356

Morinaga T (1954) Classification of rice varieties on the basis of affinity. In: International Rice Commission, working party on rice breeding. Rep 5th meeting, pp 1–14

Morinaga T (1964) Cytogenetical investigations on *Oryza* species. In: Tsunoda S, Takahashi N (eds) Rice genetics and cytogenetics. Elsevier, Amsterdam, pp 91–102

Morishima H, Sano Y, Oka HI (1992) Evolutionary studies in cultivated rice and its wild relatives. In: Douglas JA, Futuyma J (eds) Oxford surveys in evolutionary biology. Oxford University Press, Oxford

Mullins M, Hilu W (2002) Sequence variation in the gene encoding the 10-kDa prolamin in *Oryza* (Poaceae). I Phylogenetic Implications. Theor Appl Genet 105(6–7):841–846

Nishikawa T, Vaughan DA, Kadowaki K (2005) Phylogenetic analysis of *Oryza* species, based on simple sequence repeats and their flanking nucleotide sequences from the mitochondrial and chloroplast genomes. Theor Appl Genet 110(4):696–705

Oka HI (1958) Intervarietal variation and classification of cultivated rice. Indian J Genet Plant Breed 18:79–89

Oryzabase (n.d.). https://shigen.nig.ac.jp/rice/Oryzabase/

Rakshit S, Rakshit A, Matsumura H, Takahashi Y, Hasegawa Y, Ito A, Ishii T, Miyashita NT, Terauchi R (2007) Large-scale DNA polymorphism study of *Oryza sativa* and *O. rufipogon* reveals the origin and divergence of Asian rice. Theor Appl Genet 114(4):731–743

Ramos-Madrigal J, Smith BD, Moreno-Mayar JV, Gopalakrishnan S, Ross-Ibarra J, Gilbert MTP, Wales N (2016) Genome sequence of a 5,310-year-old maize cob provides insights into the early stages of maize domestication. Curr Biol 26(23):3195–3201

Sakai H, Ikawa H, Tanaka T, Numa H, Minami H, Fujisawa M, Shibata M, Kurita K, Kikuta A, Hamada M, Kanamori H, Namiki N, Wu J, Itoh T, Matsumoto T, Sasaki T (2011) Distinct evolutionary patterns of *Oryza glaberrima* deciphered by genome sequencing and comparative analysis. Plant J 66(5):796–805

Sakai H, Kanamori H, Arai-Kichise Y, Shibata-Hatta M, Ebana K, Oono Y, Kurita K, Fujisawa H, Katagiri S, Mukai Y, Hamada M, Itoh T, Matsumoto T, Katayose Y, Wakasa K, Yano M, Wu J (2014) Construction of pseudomolecule sequences of the *aus* rice cultivar Kasalath for comparative genomics of Asian cultivated rice. DNA Res 21(4):397–405

Sang T, Ge S (2007) Genetics and phylogenetics of rice domestication. Curr Opin Genet Dev 17(6):533–538

Sang T, Ge S (2013) Understanding rice domestication and implications for cultivar improvement. Curr Opin Plant Biol 16(2):139–146

Schatz MC, Maron LG, Stein JC, Hernandez Wences A, Gurtowski J, Biggers E, Lee H, Kramer M, Antoniou E, Ghiban E, Wright MH, Chia JM, Ware D, McCouch SR, McCombie WR (2014) Whole genome de novo assemblies of three divergent strains of rice, *Oryza sativa*, document novel gene space of *aus* and *indica*. Genome Biol 15(11):506

Scott MF, Botigué LR, Brace S, Stevens CJ, Mullin VE, Stevenson A, Thomas MG, Fuller DQ, Mott R (2019) A 3,000-year-old Egyptian emmer wheat genome reveals dispersal and domestication history. Nat Plants 5(11):1120–1128

Shomura A, Izawa T, Ebana K, Ebitani T, Kanegae H, Konishi S, Yano M (2008) Deletion in a gene associated with grain size increased yields during rice domestication. Nat Genet 40(8):1023–1028

Stein JC, Yu Y, Copetti D, Zwickl DJ, Zhang L, Zhang C, Chougule K, Gao D, Iwata A, Goicoechea JL, Wei S, Wang J, Liao Y, Wang M, Jacquemin J, Becker C, Kudrna D, Zhang J, Londono CEM, Song X, Lee S, Sanchez P, Zuccolo A, Ammiraju JSS, Talag J, Danowitz A, Rivera LF, Gschwend AR, Noutsos C, Wu CC, Kao SM, Zeng JW, Wei FJ, Zhao Q, Feng Q, El Baidouri M, Carpentier MC, Lasserre E, Cooke R, Rosa Farias DD, da Maia LC, Dos Santos RS, Nyberg KG, McNally KL, Mauleon R, Alexandrov N, Schmutz J, Flowers D, Fan C, Weigel D, Jena KK, Wicker T, Chen M, Han B, Henry R, Hsing YC, Kurata N, de Oliveira AC, Panaud O, Jackson SA, Machado CA, Sanderson MJ, Long M, Ware D, Wing RA (2018) Genomes of 13 domesticated and wild rice relatives highlight genetic conservation, turnover and innovation across the genus *Oryza*. Nat Genet 50(2):285–296

Tanaka K, Kamijo N, Tabuchi H, Hanamori K, Matsuda R, Suginomori J, Sato YI, Udatsu T, Ishikawa R (2016) Morphological and molecular genetics of ancient remains and modern rice (*Oryza sativa*) confirm diversity in ancient Japan. Genet Resour Crop Evol 63:447–464

Tanaka K, Zhao C, Wang N, Kubota S, Kanehara M, Kamijo N, Ishikawa R, Tasaki H, Kanehara M, Liu B, Chen M, Nakamura S, Udatsu T, Wang C (2020a) Classification of archaic rice grains excavated at the Mojiaoshan site within the Liangzhu site complex reveals an *Indica* and *Japonica* chloroplast complex. Product Process Nutr 2:15

Tanaka N, Shenton M, Kawahara Y, Kumagai M, Sakai H, Kanamori H, Yonemaru J, Fukuoka S, Sugimoto K, Ishimoto M, Wu J, Ebana K (2020b) Whole-genome sequencing of the NARO world rice core collection (WRC) as the basis for diversity and association studies. Plant Cell Physiol 61(5):922–932

Tanaka N, Shenton M, Kawahara Y, Kumagai M, Sakai H, Kanamori H, Yonemaru JI, Fukuoka S, Sugimoto K, Ishimoto M, Wu J, Ebana K (2021) Investigation of the genetic diversity of a rice core collection of Japanese landraces using whole-genome sequencing. Plant Cell Physiol 61(12):2087–2096

Tatsumi J, Kawano K (1972) Suitou no noge ni tsuite [About the awn in rice]. Res Rep Tokai Branch Crop Sci Soc Japan 56:11–15. (In Japanese)

Trucchi E, Benazzo A, Lari M, Iob A, Vai S, Nanni L, Bellucci E, Bitocchi E, Raffini F, Xu C, Jackson SA, Lema V, Babot P, Oliszewski N, Gil A, Neme G, Michieli CT, De Lorenzi M, Calcagnile L, Caramelli D, Star B, de Boer H, Boessenkool S, Papa R, Bertorelle G (2021) Ancient genomes reveal early Andean farmers selected common beans while preserving diversity. Nat Plants 7(2):123–128

Vallebueno-Estrada M, Rodríguez-Arévalo I, Rougon-Cardoso A, Martínez González J, García Cook A, Montiel R, Vielle-Calzada JP (2016) The earliest maize from San Marcos Tehuacán is a partial domesticate with genomic evidence of inbreeding. Proc Natl Acad Sci U S A 113(49):14151–14156

Vaughan DA (1994) The wild relatives of rice: a genetic resources handbook. International Rice Reseach Institute, Philippines

Vitte C, Ishii T, Lamy F, Brar D, Panaud O (2004) Genomic paleontology provides evidence for two distinct origins of Asian rice (*Oryza sativa* L.). Mol Gen Genomics 272(5):504–511

Wang ZY, Second G, Tanksley SD (1992) Polymorphism and phylogenetic relationships among species in the genus *Oryza* as determined by analysis of nuclear RFLPs. Theor Appl Genet 83(5):565–581

Wang H, Vieira FG, Crawford JE, Chu C, Nielsen R (2017) Asian wild rice is a hybrid swarm with extensive gene flow and feralization from domesticated rice. Genome Res 27(6):1029–1038

Wang W, Mauleon R, Hu Z, Chebotarov D, Tai S, Wu Z, Li M, Zheng T, Fuentes RR, Zhang F, Mansueto L, Copetti D, Sanciangco M, Palis KC, Xu J, Sun C, Fu B, Zhang H, Gao Y, Zhao X, Shen F, Cui X, Yu H, Li Z, Chen M, Detras J, Zhou Y, Zhang X, Zhao Y, Kudrna D, Wang C, Li R, Jia B, Lu J, He X, Dong Z, Xu J, Li Y, Wang M, Shi J, Li J, Zhang D, Lee S, Hu W, Poliakov A, Dubchak I, Ulat VJ, Borja FN, Mendoza JR, Ali J, Li J, Gao Q, Niu Y, Yue Z, Naredo MEB, Talag J, Wang X, Li J, Fang X, Yin Y, Glaszmann JC, Zhang J, Li J, Hamilton RS, Wing RA, Ruan J, Zhang G, Wei C, Alexandrov N, McNally KL, Li Z, Leung H (2018) Genomic variation in 3,010 diverse accessions of Asian cultivated rice. Nature 557(7703):43–49

Xu X, Liu X, Ge S, Jensen JD, Hu F, Li X, Dong Y, Gutenkunst RN, Fang L, Huang L, Li J, He W, Zhang G, Zheng X, Zhang F, Li Y, Yu C, Kristiansen K, Zhang X, Wang J, Wright M, McCouch S, Nielsen R, Wang J, Wang W (2011) Resequencing 50 accessions of cultivated and wild rice yields markers for identifying agronomically important genes. Nat Biotechnol 30(1):105–111

Yang CC, Kawahara Y, Mizuno H, Wu J, Matsumoto T, Itoh T (2012) Independent domestication of Asian rice followed by gene flow from *japonica* to *indica*. Mol Biol Evol 29(5):1471–1479

Yano K, Yamamoto E, Aya K, Takeuchi H, Lo PC, Hu L, Yamasaki M, Yoshida S, Kitano H, Hirano K, Matsuoka M (2016) Genome-wide association study using whole-genome sequencing rapidly identifies new genes influencing agronomic traits in rice. Nat Genet 48(8):927–934

Zhang J, Chen LL, Sun S, Kudrna D, Copetti D, Li W, Mu T, Jiao WB, Xing F, Lee S, Talag J, Song JM, Du B, Xie W, Luo M, Maldonado CE, Goicoechea JL, Xiong L, Wu C, Xing Y, Zhou DX, Yu S, Zhao Y, Wang G, Yu Y, Luo Y, Hurtado BE, Danowitz A, Wing RA, Zhang Q (2016a) Building two *indica* rice reference genomes with PacBio long-read and Illumina paired-end sequencing data. Sci Data 3:160076

Zhang J, Chen LL, Xing F, Kudrna DA, Yao W, Copetti D, Mu T, Li W, Song JM, Xie W, Lee S, Talag J, Shao L, An Y, Zhang CL, Ouyang Y, Sun S, Jiao WB, Lv F, Du B, Luo M, Maldonado CE, Goicoechea JL, Xiong L, Wu C, Xing Y, Zhou DX, Yu S, Zhao Y, Wang G, Yu Y, Luo Y, Zhou ZW, Hurtado BE, Danowitz A, Wing RA, Zhang Q (2016b) Extensive sequence divergence between the reference genomes of two elite *indica* rice varieties Zhenshan 97 and Minghui 63. Proc Natl Acad Sci U S A 113(35):E5163–E5171

Zhao K, Wright M, Kimball J, Eizenga G, McClung A, Kovach M, Tyagi W, Ali ML, Tung CW, Reynolds A, Bustamante CD, McCouch SR (2010) Genomic diversity and introgression in *O. sativa* reveal the impact of domestication and breeding on the rice genome. PLoS One 5(5):e10780

Zhao K, Tung CW, Eizenga GC, Wright MH, Ali ML, Price AH, Norton GJ, Islam MR, Reynolds A, Mezey J, McClung AM, Bustamante CD, McCouch SR (2011) Genome-wide association mapping reveals a rich genetic architecture of complex traits in *Oryza sativa*. Nat Commun 2:467
Zhou Y, Chebotarov D, Kudrna D, Llaca V, Lee S, Rajasekar S, Mohammed N, Al-Bader N, Sobel-Sorenson C, Parakkal P, Arbelaez LJ, Franco N, Alexandrov N, Hamilton NRS, Leung H, Mauleon R, Lorieux M, Zuccolo A, McNally K, Zhang J, Wing RA (2020) A platinum standard pan-genome resource that represents the population structure of Asian rice. Sci Data 7(1):113
Zhu Q, Ge S (2005) Phylogenetic relationships among A-genome species of the genus *Oryza* revealed by intron sequences of four nuclear genes. New Phytol 167(1):249–265
Zou XH, Zhang FM, Zhang JG, Zang LL, Tang L, Wang J, Sang T, Ge S (2008) Analysis of 142 genes resolves the rapid diversification of the rice genus. Genome Biol 9(3):R49

Chapter 10
Wheat: Recent Advances in Genomics to Explore the History of Domestication and Polyploidization

Moeko Okada and Kentaro K. Shimizu

Abstract Wheat is cultivated in the largest area among the crop species. The geographic spread of wheat since its domestication in the Middle East approximately 10,000 years ago has attracted a broad interest in biology and archaeology. The spread of wheat associated with the migration of farmers and languages has been proposed. However, the complexity of the wheat genome due to polyploidization has been an obstacle in studying the domestication and spread of wheat using genetic data. In this chapter, we introduce domesticated wheat species in the context of polyploidization and then summarize recent genomic studies on the independent domestication events of three wheat species: Einkorn, Emmer, and Timopheevi. Population genetic studies supported the origin place of Einkorn and Emmer in the northern part of the Fertile Crescent and the importance of gene flow. Approximately 8000 years ago, common wheat emerged through a further polyploidization event, followed by its spread westward to Europe and eastward to East Asia, including Japan. We discuss the potential of growing genomic data, together with archaeobotanical studies and

M. Okada (✉)
Department of Evolutionary Biology and Environmental Studies, University of Zurich, Zurich, Switzerland

Kihara Institute for Biological Research, Yokohama City University, Yokohama, Kanagawa, Japan

Graduate School of Science and Technology, Niigata University, Niigata, Japan
e-mail: mookad@agr.niigata-u.ac.jp

K. K. Shimizu
Department of Evolutionary Biology and Environmental Studies, University of Zurich, Zurich, Switzerland

Kihara Institute for Biological Research, Yokohama City University, Yokohama, Kanagawa, Japan
e-mail: kentaro.shimizu@uzh.ch

N. Osada et al. (eds.), *Phylogeographic History of Plants and Animals Coexisting with Humans in Asia*, Evolutionary Studies,
https://doi.org/10.1007/978-981-97-6887-5_10

ancient or historical genomics, to reveal the longstanding question of wheat domestication and spread.

Keywords Wheat · Allopolyploidization · Domestication · Genomics · Sequencing

Abbreviations

aDNA	Ancient DNA
SNP	single nucleotide polymorphism

10.1 Introduction

The Neolithic period, which started around 10,000 years ago, was characterized by agriculture (Smith 1998). Crop species have agronomic and economic value as crucial resources for food, clothing, medicines, and other products. Wheat, defined as plants belonging to the genus *Triticum*, was independently domesticated at least three times. Today, wheat is the most widely grown crop (FAOstat 2022) and is mainly used as food and feed for domesticated animals. Two wheat species, i.e., common wheat, also called bread wheat, and durum wheat, have spread worldwide by demic and cultural diffusion (Stoneking 2017) since their domestication and now support approximately 1/5 of the global calorie and protein consumption in the human diet (Shewry and Hey 2015). Common wheat (*Triticum aestivum* L., $2n = 6x = 42$, AABBDD genome, also called bread wheat) is a hexaploid mainly used to make bread, noodles, and confectioneries (Mastrangelo and Cattivelli 2021). Durum wheat (*Triticum turgidum* ssp. *durum* Desf., $2n = 4x = 28$, AABB genome) is tetraploid and is mainly used to make pasta or couscous (Mastrangelo and Cattivelli 2021). Common wheat accounts for approximately 95% of the world's current wheat production (Mastrangelo and Cattivelli 2021; Gutaker and Purugganan 2023). During wheat domestication and subsequent breeding processes including the Green Revolution, agricultural traits (e.g., resistance to biotic and abiotic stresses), taste, and traits related to the local food culture have been under selection (Levy and Feldman 2022). In contrast to other crops domesticated from their wild ancestors without changing the ploidy levels (e.g., rice, barley, and maize), wheat provides insights into domestication and diversification through polyploidization.

Although wheat is one of the most important crops and many wheat seeds have been recovered from archaeological sites, ancient and modern genomics of wheat lags behind those of other crops, such as rice and maize. This is partly due to the large size and complexity of the allopolyploid genome of wheat derived from interspecific hybridization and chromosomal doubling, and the highly repetitive sequences (The International Wheat Genome Sequence Consortium (IWGSC) 2018). Recent improvements in next-generation sequencing, analytical methods

(Kuo et al. 2018, 2020; Nishimura et al. 2022; Sun et al. 2023), and chromosome-scale genome assemblies (IWGSC 2018; Walkowiak et al. 2020) have made both modern and ancient genomics feasible. This chapter begins with an outline of wheat allopolyploid evolution. Next, we summarize the domestication events of three independent species—Einkorn, Emmer, and Timopheevi—and then explain their diversification after domestication. Finally, we provide an overview of recent advances in modern genomics of wheat that facilitate further archaeological and evolutionary studies on wheat.

10.2 Allopolyploidization and Domestication in Wheat Relatives

Domesticated wheat species, as well as some of their relatives that belong to the genera *Triticum* and *Aegilops*, encompassing diploid (2*x*), tetraploid (4*x*), and hexaploid (6*x*) species are shown in Fig. 10.1. The allopolyploid species in wheat (4*x* and 6*x*) were formed by allopolyploidization (i.e., genome duplication with hybridization between species). Although the relationship between domestication and allopolyploidization is complex, it can be understood as a combination of the two scenarios commonly observed in polyploid crop species (Akagi et al. 2022). First, domestication occurred at the diploid and polyploid levels, and eventually, the polyploid species became more widely used than the diploid species. In wheat, domestication occurred once at the diploid level (*Triticum monococcum* ssp. *monococcum* L., $2n = 2x = 14$, A^mA^m genome, as the domesticated form) and twice at the tetraploid level (domesticated Emmer, *Triticum turgidum* ssp. *dicoccum* (Schrenk ex Schübl.) Tell., $2n = 4x = 28$, AABB genome, as the domesticated form; and domesticated Timopheevi, *Triticum timopheevii* ssp. *timopheevii* Zhuk., $2n = 4x = 28$, AAGG genome, as the domesticated form). Among these three domesticated species, only tetraploid *T. turgidum* is widely cultivated today, as aforementioned. Second, after a species is domesticated, subsequent allopolyploidization events often contribute to crop improvement. The most widely cultivated common wheat is a hexaploid species obtained through "natural hybridization" (non-intended interspecific and intraspecific crosses) between domesticated tetraploid wheat and wild diploid species.

Two wild tetraploid species—wild Emmer wheat, *Triticum turgidum* ssp. *dicoccoides* (Asch. & Graebn.) Schweinf., ($2n = 2x = 28$, AABB) and wild Timopheevi, *Triticum timopheevii* (Zhuk.) Zhuk. ssp. *armeniacum* (Jakubz.) Slageren (Syn. *T. araraticum* Jakubz. $2n = 4x = 28$, AAGG)—were estimated to have emerged less than 0.5 million years ago without the influence of domestication (Fig. 10.1). In both tetraploid species, one of the diploid progenitors is *Triticum urartu* Tumanian ex Gandilyan ($2n = 2x = 14$, AA genome). The nature of the ancestral species of the B and G genomes has long been discussed; however, recent genome analyses suggest that both are *Aegilops speltoides* Tausch ($2n = 2x = 14$, SS) or a species closely related to it but currently non-existent (Sarkar and Stebbins 1956; Dvořák 1976;

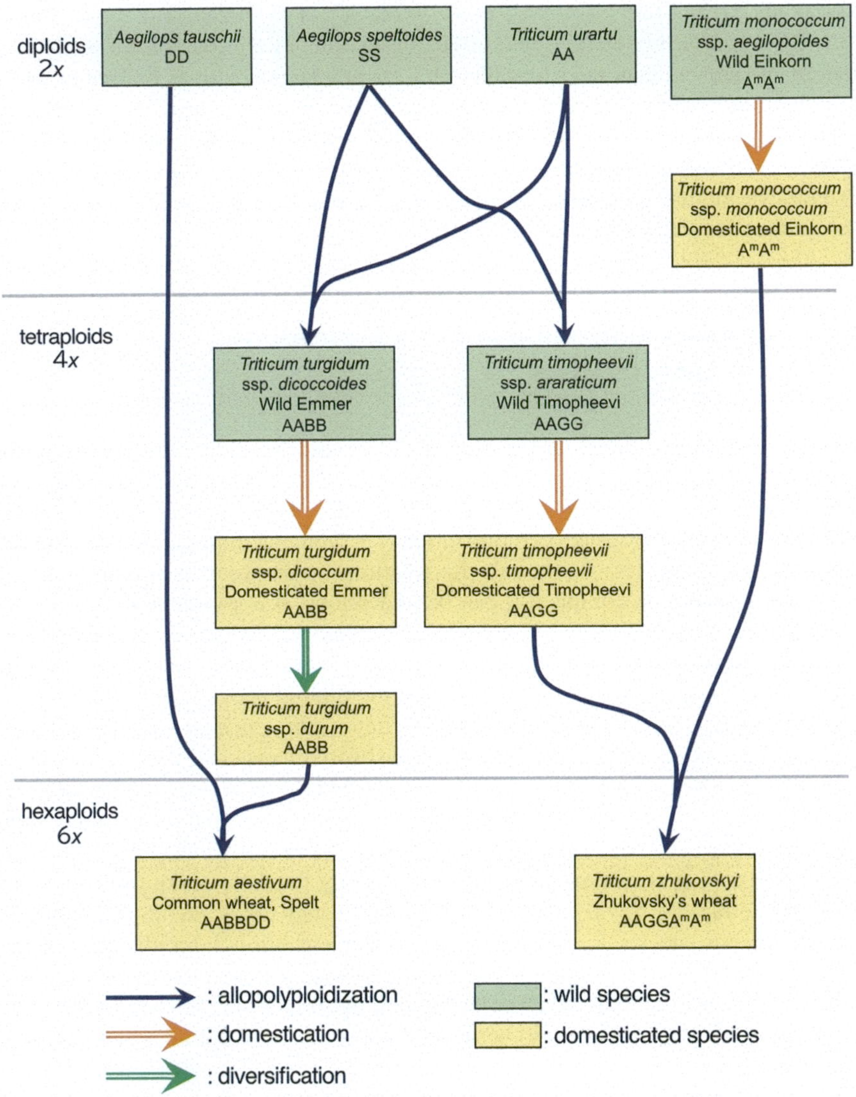

Fig. 10.1 Domesticated wheat species, its ancestral species, and allopolyploidization events. Dark blue, orange and green arrows represent allopolyploidization, domestication, and diversification, respectively. Green and yellow boxes indicate wild and domesticated species, respectively

Chapman et al. 1976; Dvorák et al. 1988, 1993; Ogihara and Tsunewaki 1988; Noda and Ge 1989; Dvorák and Zhang 1990; Miyashita et al. 1994; Huang et al. 2002; Kilian et al. 2007a; Matsuoka 2011; Li et al. 2022; Avni et al. 2022).

Both these two wild tetraploid species, wild Emmer and wild Timopheevi, were domesticated to form the domesticated Emmer and the domesticated Timopheevi

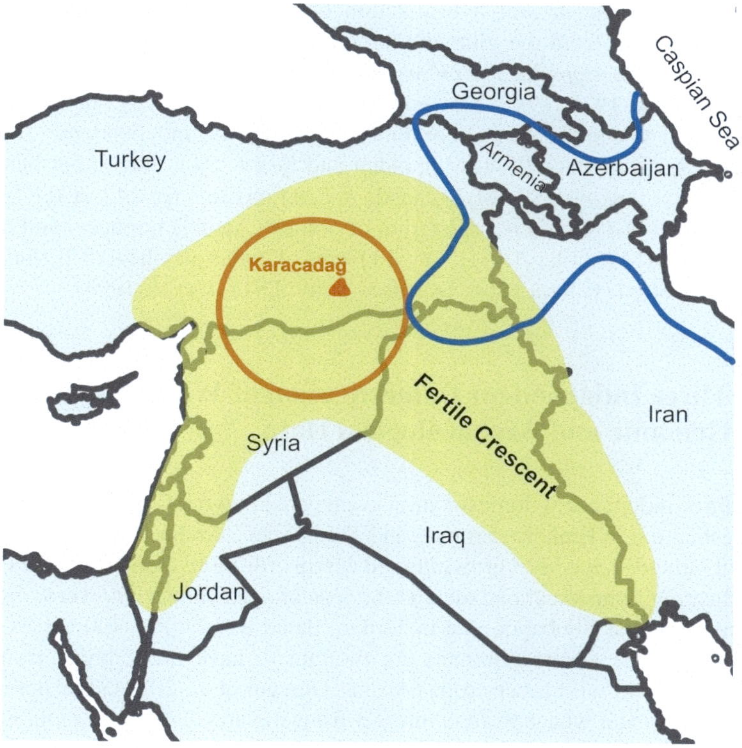

Fig. 10.2 Schematic ranges relevant to the domestication place of wheat based on a free shape file downloaded from http://www.naturalearthdata.com/ and made using R package "ggplot2" ver. 3.5.0 in R software ver. 4.3.2. The Fertile Crescent is shown in yellow. The red circle indicates the 'core area' where agriculture is thought to have emerged, which includes the Karacadağ region, as shown by the red triangle. The blue line indicates the part of the possible area where natural hybridization between domesticated tetraploid wheat and *Ae. tauschii* may have generated common wheat. (Modified from Lev-Yadun et al. 2000; Matsuoka 2011). Countries mentioned in the text are denoted

(Fig. 10.1; Feldman 2001; Salamini et al. 2002). From domesticated Emmer wheat, further selection led to free-threshing durum wheat (*T. turgidum* ssp. *durum* (Desf.) MacKey) and several other tetraploid wheat species (Feldman and Kislev 2007; Luo et al. 2007; Matsuoka 2011); simultaneously, from wild diploid Einkorn wheat (*T. monococcum* ssp. *aegilopoides*, Syn. *T. boeoticum* Boiss.), domesticated diploid Einkorn wheat (*T. monococcum* ssp. *monococcum* L.) was formed. The domestications of Einkorn and Emmer are thought to have occurred in the Fertile Crescent region (yellow shaded area in Fig. 10.2). Little is known about the domestication of Timopheevi (described below) than that about Emmer (Czajkowska et al. 2020; Badaeva et al. 2022).

After these domestication events, another round of hybridization occurred, and two hexaploid species were created independently (Fig. 10.1). One is the common wheat formed by the natural hybridization between a free-threshing (i.e., easy release of the grain from the glumes), tetraploid wheat, *T. turgidum* and the wild diploid relative, *Aegilops tauschii* Coss. ($2n = 2x = 14$, DD) (Syn. *Ae. squarrosa* Eig.) (Muramatsu 1979, 1985; Matsuoka 2011; Pont et al. 2019). *Ae. tauschii* has been identified as a progenitor of common wheat and was independently found by Kihara and Sears (Kihara 1944; McFadden and Sears 1944). The other hexaploid wheat, *Triticum zhukovskyi* A. M. Menabde & Eritzjan ($2n = 6x = 42$, $AAGGA^mA^m$), was obtained by the natural hybridization of domesticated Timopheevi and domesticated Einkorn (Matsuoka 2011). No wild forms of common wheat or *T. zhukovskyi* have been reported (Kihara 1966; Dorofeev 1969; Dvorák et al. 1993).

10.3 Three Independent Domestication of Wheat Studied by Genomic and Archaeological Data

At least three independent domestication events are known among *Triticum* species: the domestication of Einkorn, Emmer, and Timopheevi. Archaeobotanists generally agree that humans have been growing wild cereals for more than 1000 years before traits indicating domestication, such as the loss of seed shattering, were observed. For example, at the Shubayqa site in Jordan, dated to 12,400–9400 cal BC (calibrated year Before Christ), humans are thought to have made bread from wild Einkorn wheat and wild tuber crops (Arranz-Otaegui et al. 2018). The domestication periods of each wheat were estimated from the archaeological remains: wild Einkorn was domesticated first, and ~1000 years later, wild Emmer was domesticated. Barley and rye were domesticated during the same period in the Fertile Crescent (Nesbitt 2001; Fuller 2007; Kuijt and Finlayson 2009; Willcox 2012; Riehl et al. 2013). However, where and how many times these species were domesticated has long discussed. Answering these questions is pivotal to our understanding of the Neolithic transition, which marked a shift from a hunter-gatherer to farming lifestyle. This section will describe each domestication event of the three *Triticum* species: Einkorn, Emmer, and Thimopheevi.

Domestication of Wild Einkorn

Einkorn was the first wheat species domesticated by human beings ~10,000 years ago. It exhibits a hulled phenotype, in which releasing the grain from the glumes is difficult. Population genetic analysis showed that wild Einkorn has three races (lineages): α, β, and γ (Ahmed et al. 2023). Races α and β were distributed only in restricted areas around the Karacadağ region (red triangle in Fig. 10.3) within the Fertile Crescent. Race γ was not found in the Fertile Crescent and was distributed mainly in central and western Turkey (Fig. 10.3).

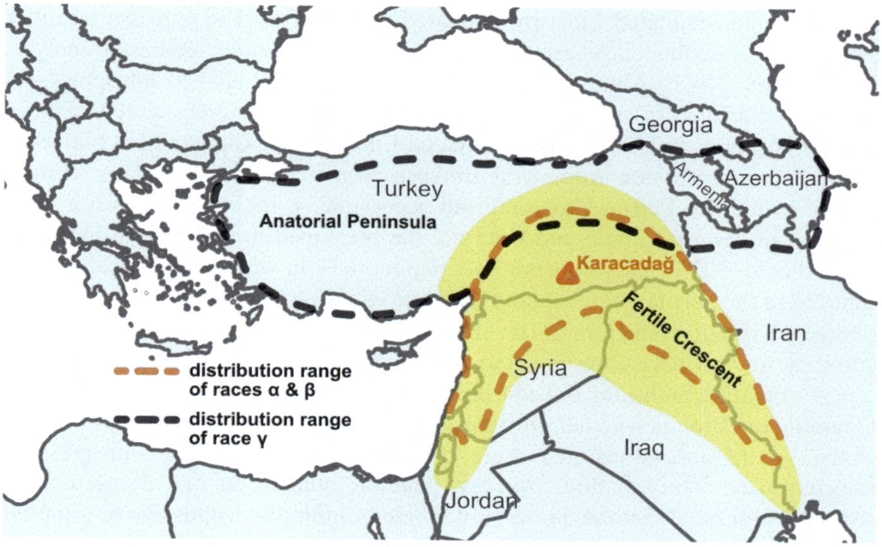

Fig. 10.3 Schematic ranges relevant to the domestication of Einkorn. The main distribution area of wild Einkorn races is depicted, with the Fertile Crescent highlighted in yellow. The red dashed line represents the distribution areas of races α and β, while the black dashed line represents the main distribution area of race γ. Countries mentioned in the text are denoted. The map was created by the same methods as Fig. 10.2

Based on archaeological remains, two potential domestication locations of Einkorn were suggested: the Fertile Crescent (red dashed line in Fig. 10.3) and central and western Turkey in the Anatolian Peninsula, outside the Fertile Crescent (black dashed line in Fig. 10.3; Zohary and Hopf 1993, 2000). Heun et al. (1997) studied 338 accessions of wild and domesticated Einkorn from the Fertile Crescent and Europe using 228 amplified fragment length polymorphisms (AFLPs) markers located on non-coding sequences. They demonstrated that wild Einkorn accessions were most closely associated with domesticated Einkorn accessions originating from the Karacadağ region, which is in the fertile crescent (red triangle in Fig. 10.3). This finding is consistent with the origin in the Fertile Crescent, but it was insufficient to draw a strong conclusion on the domestication location of Einkorn because of the limited amount of genetic information used in this study.

Recent genomics has revealed the details of Einkorn domestication. Ahmed et al. (2023) used 219 Einkorn accessions with newly assembled genome sequences of wild and domesticated Einkorn. The domesticated Einkorn accessions clustered together with race β, most of which were collected from the area adjacent to the Karacadağ region in the Fertile Crescent (red triangle in Fig. 10.3). These findings support previous reports and the hypothesis that Einkorn underwent domestication from a limited and confined wild population closely linked to β accessions. Although nucleotide diversity is typically reduced by bottlenecks in most domesticated species, no significant reduction in nucleotide diversity was detected in domesticated Einkorn compared with that of wild Einkorn in either genotyping-by-sequencing-based analysis or reference sequence-based analysis (Kilian et al. 2007b; Ahmed et al. 2023). This was attributed to introgressions from the wild Einkorn race γ to an average of 6.7% of the domesticated Einkorn genome (Ahmed et al. 2023). Domesticated Einkorn accessions with a high proportion of introgression from wild Einkorn race γ were collected in western Turkey within the Fertile Crescent, and accessions without introgression were collected in eastern Turkey and Georgia, the presumed domestication locations (Ahmed et al. 2023). This suggests two-step models, in which domestication first occurred in the Fertile Crescent, and then introgression from another lineage contributed to the diversification. The results also suggest that introgression from wild Einkorn may have contributed to the high nucleotide diversity and adaptation to new climatic conditions outside the Fertile Crescent (Ahmed et al. 2023). A remaining question is whether introgression occurred in wild species, resulting in admixed wild accessions that were domesticated, or whether introgression occurred after domestication. Notably, genomic studies on rice domestication have also suggested that the question of single or multiple origins can be simplistic and may depend on the definition of domestication and origin (Molina et al. 2011; Choi et al. 2017).

Domestication of Wild Emmer and the Emergence of Hexaploid Common Wheat

T. turgidum, including durum wheat used mainly for pasta and couscous, was domesticated from wild Emmer wheat (Fig. 10.1). Early genetic studies have suggested that wild Emmer has two subpopulations: the southern population that grows in the Southern Levant (green line in Fig. 10.4), and the northern population that grows in southeastern Turkey, northern Iraq, and northwestern Iran, so-called Northern Levant (blue line in Fig. 10.4; Ozkan et al. 2002). Whether the wild Emmer population from the north or south contributes to domestication has long been debated (Fig. 10.4).

The remains of domesticated Emmer were found at several pre-pottery Neolithic sites in the Fertile Crescent region, both in the southern and northern parts. The oldest remains were from an archaeological site dating ~9800–9300 years ago in Tell Aswad, Syria (red dot in Fig. 10.4; Willcox 2012). Domesticated Emmer appeared almost simultaneously in the southern and northern parts of the Fertile Crescent 9500–9000 years ago (Feldman and Kislev 2007). Possible hypotheses are (1) independent domestication in the southern and northern parts and (2) the introduction from the northern to the southern parts immediately after domestication, or vice versa. The domestication site of wild Emmer has been investigated using molecular marker analyses based on chloroplast DNA haplotypes (Mori et al. 2003), nuclear DNA analysis using restriction fragment length polymorphisms (RFLP) (Luo et al. 2007), and AFLP (Ozkan et al. 2005). All

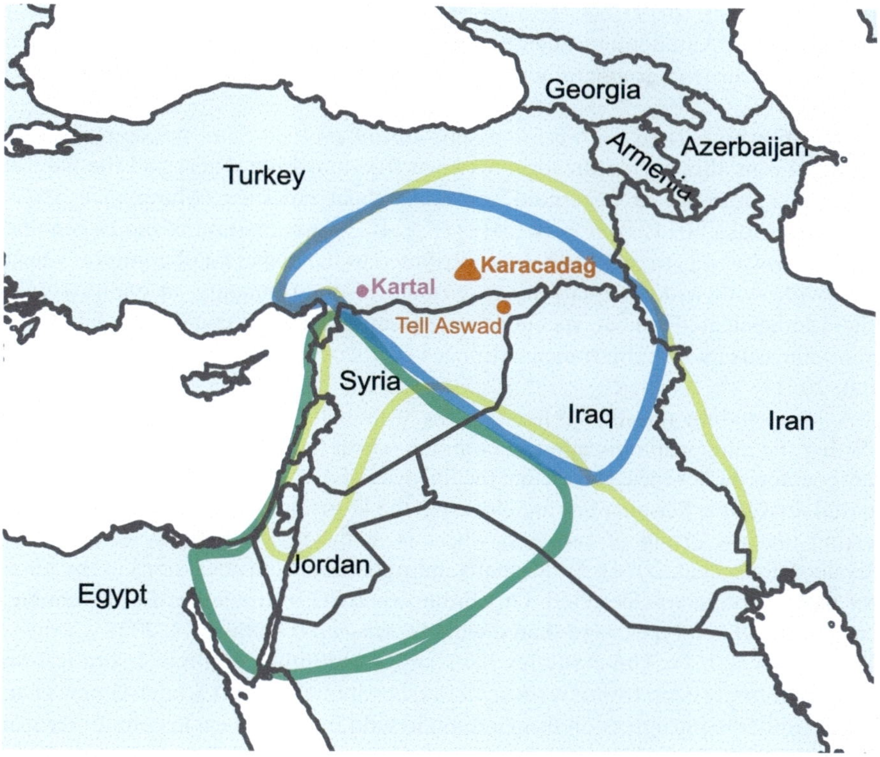

Fig. 10.4 Schematic candidate area of emmer domestication. The Fertile Crescent is shown by the yellow line. The blue and green lines represent the northern and southern Levant areas, respectively. Countries mentioned in the text are denoted. The map was created by the same methods as Fig. 10.2

these studies have shown that the northern population distributed in the blue-lined area of Fig. 10.4 played an important role in the domestication of wild Emmer, particularly around the Karacadağ region (red triangle in Fig. 10.4) or the Kartal region (pink dot in Fig. 10.4). Additionally, Civáň et al. (2013) supported through nuclear DNA polymorphisms of LTR retrotransposon insertion in the rRNA gene that the wild populations from the Karacadağ region (red triangle in Fig. 10.4) were the place of domesticated Emmer wheat.

The release of the assembled genome sequence of the wild Emmer, Zavitan (Avni et al. 2017), further advanced the study of Emmer wheat domestication. Population genetic analysis using whole genome resequencing data of 34 wild and 31 domesticated Emmer accessions and Zavitan's reference genome sequence revealed that the diffusion of domesticated Emmer accessions originated from a single site in the Northern Levant (blue line in Fig. 10.4), particularly in southeastern Turkey close to the Karacadağ region in the Fertile Crescent (red triangle in Fig. 10.4), which supported a putative domesticated location there (Civáň et al.

2013; Avni et al. 2017) (see section "Morphological Changes in Domesticated Wheat: Loss of Autonomous Seed Shattering"). Genomic studies have also shown gene flow from the southern wild Emmer to the domesticated Emmer (Zhao et al. 2023).

Free-threshing *Triticum turgidum* spp. *durum* evolved from the domesticated Emmer. From allopolyploidization between free-threshing wheat and the natural species *Ae. tauschii*, a hexaploid common wheat emerged (Muramatsu 1979, 1985; Matsuoka 2011; Pont et al. 2019) (Fig. 10.1). Acquisition of the D-genome from *Ae. tauschii* is thought to have contributed to the dispersal of common wheat worldwide with wide adaptability to a range of environments, in contrast with other domesticated wheat species that are cultivated in restricted regions in Transcaucasus and Mediterranean climates (Dubcovsky and Dvorák 2007; Asuke et al. 2021).

A longstanding question related to the establishment of hexaploid wheat is whether the allohexaploidization of common wheat has single or multiple origins. Pan-genome studies have shown that the lineages of *Ae. tauschii* in the southwestern coastal area of the Caspian Sea are closest to the D-genome of common wheat, suggesting that the origin of common wheat is in the region (Gaurav et al. 2022; Cavalet-Giorsa et al. 2024). Additionally, introgression from accessions of *Ae. tauschii* from Georgia was detected. On chromosome 7D of the wheat D subgenome, four different haplotypes were shared with *Ae. tauschii* (Gaurav et al. 2022; Cavalet-Giorsa et al. 2024). These studies have suggested either multiple hybridization events or introgression from *Ae. tauschii* into hexaploid wheat (Cavalet-Giorsa et al. 2024), similar to introgression from tetraploid wild Emmer wheat to common wheat (He et al. 2019). Archaeological data suggest that common wheat emerged approximately 8000 years ago (Matsuoka 2011). Further research using genomic data is required (Zhao et al. 2023).

Domestication of Wild Timopheevi

Wild Timopheevi, another wild tetraploid species, emerged from the allopolyploidization between *T. urartu* and a species close to *Ae. speltoides*, as previously described. Genotyping-by-sequencing analysis suggested that northern Iraq and northwestern Iran (green circle in Fig. 10.5) are the origin of the allotetraploidization of wild Timopheevi (Yadav et al. 2023). The domesticated form of wild Timopheevi is a domesticated hulled Timopheevi wheat (*T. timopheevii* ssp. *timopheevii*). Domesticated Timopheevi shows morphological characteristics similar to those of *T. turgidum*. Additionally, domesticated Timopheevi can form hybrids with *T. turgidum*; however, the F_1 hybrid is sterile because of chromosomal differences, suggesting that these two species are genetically distinct (Kihara and Lilienfeld 1934; Svetozarova 1939).

The domestication location of Timopheevi has long been investigated using archaeological methods and chloroplast genome analysis; however, no conclusions have been reached. Recent genomic studies have provided novel insights. Wild

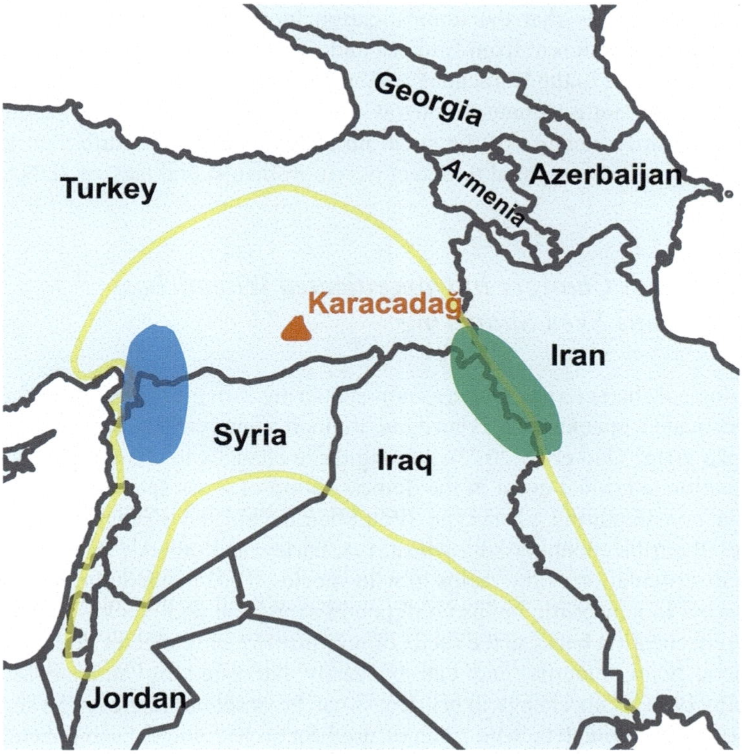

Fig. 10.5 Schematic candidate area of Timopheevi domestication. The Fertile Crescent is shown in yellow. The blue and green circles represent the potential domestication places of Timopheevi suggested by Mori et al. (2009) and Nave et al. (2021), respectively. The putative domestication place differs from other domesticated wheat species, Einkorn and Emmer, which are supposed to have occurred close to the Karacadağ region shown in the red triangle. Countries mentioned in the text are denoted. The map was created by the same methods as Fig. 10.2

Timopheevi grow in southeastern Turkey, northern Iraq, western Iran, Armenia, and Azerbaijan (Fig. 10.5; Zohary and Hopf 1993, 2000), and domesticated Timopheevi are mainly found in western Georgia, where wild Timopheevi do not grow. These archaeological findings suggest that the domestication location of Timopheevi was not Georgia, and after domestication, Timopheevi wheat might have been introduced into Georgia.

Molecular studies have revealed detailed phylogenetic relationships between wild and domesticated Timopheevi. Chloroplast DNA polymorphism analysis performed by Mori et al. (2009) suggested that the domestication of Timopheevi is of a single origin in the region including southern Turkey and northern Syria (blue circle in Fig. 10.5). In contrast, Nave et al. (2021) conducted haplotype analysis of *Btr* genes and suggested that Iraq is a potential *T. timopheevii* domestication location (green circle in Fig. 10.5). Both findings support the archaeological results that the domestication location of Timopheevi is not in Georgia, and also different from Einkorn and Emmer wheat, which are inferred to have occurred close to the Karacadağ region shown in the red triangle in Fig. 10.5. Current data are not adequate to narrow the location of Timopheevi domestication. This is partly because of the small number of samples, insufficient number of molecular markers targeted to detect polymorphisms, and lack of aDNA data.

Morphological Changes in Domesticated Wheat: Loss of Autonomous Seed Shattering

Morphological characteristics such as non-shattering, soft glume, and free-threshing of domesticated species can contribute to increasing yields and human needs (Matsuoka 2011; Gao et al. 2017). Among these changes in characteristics during domestication, a critical event in the domestication of grass species was the selection of a non-shattering phenotype (Pourkheirandish and Komatsuda 2007). In Triticeae (the tribe encompassing wheat, rye, barley, and others), shattering occurs through disarticulation of the rachis in wild species. The change in brittleness from brittle rachis to non-brittle rachis confers non-shattering and is a key event during wheat domestication because the seeds of non-brittle rachis species do not disperse spontaneously at maturity and can be easily harvested by humans as grain-consuming crops. This change in brittleness can be observed in the rough surface at the bottom of the spikelets, which can be used for archaeological samples to distinguish domesticated wheat from wild wheat.

The brittleness of Triticeae is controlled by the genes *Non-brittle rachis 1* (*Btr1*) and *Non-brittle rachis 2* (*Btr2*). The loss-of-function mutation of either *Btr1* or *Btr2* results in the formation of the non-brittle rachis phenotype (Pourkheirandish et al. 2015; Civáň and Brown 2017). This mechanism of grain disarticulation, driven by *Btr1* and *Btr2*, is unique to the Triticeae (Sakuma et al. 2011). Because changes from brittle rachis to non-brittle rachis occurred during domestication, the alleles of *Btr* genes can be used to estimate the domestication location of wheat.

All domesticated Einkorns exhibited a non-brittle rachis phenotype, which is controlled by *Btr* genes. During the domestication of Einkorn, the substitution of a single residue (A119T) in *Btr1* located on chromosome $3A^m$ converted wild Einkorn into a non-brittle rachis (Pourkheirandish et al. 2017; Zhao et al. 2019). Population genomics of Einkorn showed that all 61 domesticated Einkorn had the same haplotype as *Btr1* (Ahmed et al. 2023). Additionally, wild Einkorn collected close to a region in Southeastern Turkey around Karacadağ mountain (red triangle in Fig. 10.2) within the fertile crescent carried the same *Btr1* and Btr2 haplotype as domesticated Einkorn (Pourkheirandish et al. 2017; Ahmed et al. 2023). These results imply that the key domestication event from brittle rachis to non-brittle rachis has a single origin in domestication, and the processes might have been a mutation in *btr1* that occurred, fixed, and spread from a region in Southeastern Turkey around

Karacadağ mountain in the Fertile Crescent (Pourkheirandish et al. 2017; Ahmed et al. 2023).

The non-brittle rachis phenotypes of the domesticated tetraploid wheat species, Emmer and durum, are also based on mutations in *Btr* genes. Due to their tetraploidy, Emmer and durum wheat have two sets of *Btr* genes (chromosome 3A: *Btr1-A* and *Btr2-A*; chromosome 3B: *Btr1-B* and *Btr2-B*) (Avni et al. 2017). Based on archaeobotanical remains, the non-shattering trait of Emmer wheat was fixed across the northern and southern Levants from 8300 years ago (Arranz-Otaegui et al. 2016; Nave et al. 2016). Comparative genomics of the domesticated durum wheat cultivar Svevo and wild Emmer accession Zavitan revealed that domesticated durum wheat has a mutation in both *Btr1* duplicated homoeologs: a 2 bp deletion in *Btr1-A* and a 4 kbp insertion in *Btr1-B* (Avni et al. 2017). Moreover, all domesticated *T. turgidum* species share a common haplotype of *Btr1-A* and *Btr1-B*, suggesting a common origin (Nave et al. 2021). In domesticated Emmer wheat, the chromosomal region harboring *Btr1-A* and *Btr1-B* showed low nucleotide diversity in diversity scans, which is consistent with selection (Avni et al. 2017; Zhou et al. 2020b). Considering the old archaeological record in the Southern Levant (see section "Domestication of Wild Emmer and the Emergence of Hexaploid Common Wheat"), domestication might have occurred independently there, but the lineage could have been extinct (Zhao et al. 2023).

10.4 Diversification After Domestication

Current durum wheat (*T. turgidum* ssp. *durum*) and other subspecies of *T. turgidum* are thought to have been derived from domesticated Emmer wheat in the eastern Mediterranean region (Feldman and Kislev 2007; Luo et al. 2007). Wild and domesticated Emmers have hard glumes and hulled grains, which are challenging to thresh but protect seeds from birds and grazers. By contrast, durum wheat and its related subspecies, namely Rivet wheat (*T. turgidum* L. ssp. *turgidum*), Polish wheat (*T. turgidum* L. ssp. *polonicum* (L.) Thell.), and Khorasan wheat (*T. turgidum* L. ssp. *turanicum* (Jakubz.) Á. Löve & D. Löve), are free-threshing with soft glumes and non-hulled seeds that facilitate harvesting. The free-threshing phenotype is a representative trait of wheat diversification. The question of whether free-threshing wheat emerges in tetraploid wheat alone or independently in both tetraploid and hexaploid wheat has attracted interest.

Muramatsu (1986) demonstrated that a free-threshing trait was acquired in domesticated Emmer after domestication. Consequently, all free-threshing tetraploid wheat varieties possess a dominant allele of the *Q* factor, located on the long arm of chromosome 5A, which arises from a gain-of-function mutation caused by a single nucleotide polymorphism in the miRNA-binding site (Sears 1954; Simons et al. 2006; Debernardi et al. 2017). Comparative sequence analysis suggests that the *Q* mutation occurred only once in both tetraploid and hexaploid wheat during the evolution of *Triticum*

(Faris et al. 2003; Simons et al. 2006). However, none of these findings definitively answer the critical question of free-threshing wheat being exclusive to tetraploid wheat or occurring independently in both tetraploid and hexaploid wheat.

Diversification after allopolyploid evolution occurred not only due to genetic mutations in the *Q* gene but also due to natural hybridization (unintended hybridization). Hybridization between wheat cultivars and wheat relatives is thought to occur in the field because of the traditional mixed cultivation of various species and ploidy levels in the Middle East and Transcaucasus (Dorofeev 1969; Matsuoka et al. 2008). This natural hybridization may have contributed to the diversification of tetraploid and hexaploid wheat.

Spelt wheat (*T. aestivum* ssp. *spelta*) is an example of natural hybridization after allopolyploidization. Spelt has two landraces of independent origin: European and Asian (Kuckuck 1959). The European spelt was once considered a progenitor for free-threshing *T. aestivum* (McFadden and Sears 1944). However, recent genetic and genomic studies have revealed that spelt is obtained by crossing free-threshing common wheat and tetraploid hulled Emmer wheat (MacKey 1966; Tsunewaki 1968; Blatter et al. 2004; Hirosawa et al. 2004; Dvorák et al. 2012).

10.5 Eastward Spread and Environmental Adaptation

After domestication and natural hybridization events, tetraploid wheat and hexaploid wheat have been spread outside of the Fertile Crescent. The spread of wheat from the Fertile Crescent westward into Europe is considered to have been "demic diffusion," that is, spread through farmers during migration. In contrast, the eastward spread is considered to have been "cultural diffusion," that is, the spread associated with the transmission of agricultural technology without human migration (Stoneking 2017).

Allopolyploidization is generally believed to contribute to the expansion of the distribution range (Dubcovsky and Dvorák 2007; Shimizu 2022). Specifically, polyploids can inherit and merge the stress responsiveness of their progenitor species, enabling them to behave similarly to generalists. Additionally, evolvability may be enhanced by mutational robustness owing to the presence of homoeologous subgenomes (Shimizu 2022). The acquisition of the D-genome might have contributed to adaptation to a wide range of environments (Dubcovsky and Dvorák 2007).

Based on archaeobotanical studies, common wheat reached Central Asia approximately 6000 years ago (Harris 2011). It branched 5200 years ago to the north toward the Altai region and 4500 years ago toward South Asia (Fig. 10.6; Liu et al. 2017; Zhou et al. 2020a). Recent archaeological findings also support that wheat arrived in the region near the Altai region between 5200 and 3200 years ago (Zhou et al. 2020a).

In Far East Asia, wheat is thought to have been introduced to the Japanese Archipelago together with rice, barley, and millet (Fujita 2013; Shoda et al. 2021). Wheat seeds estimated to be 2200 or 2000 years old have been found at archaeological sites in Japan. Since the existence and use of wheat in the Japanese Archipelago

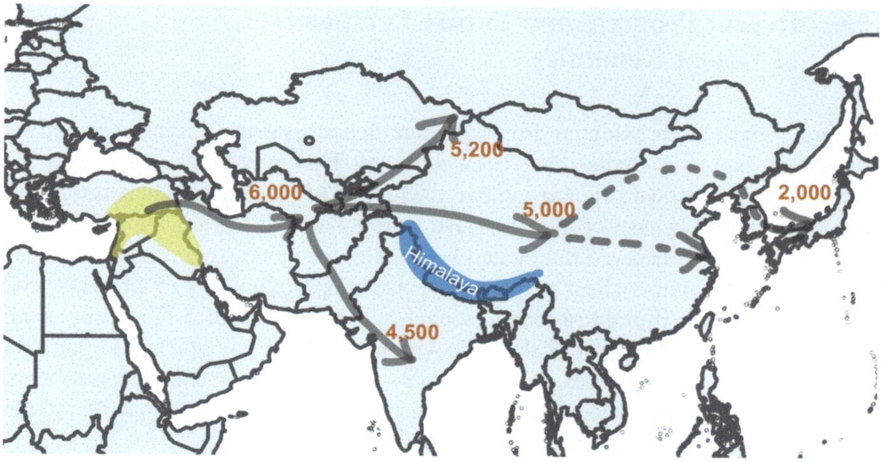

Fig. 10.6 Potential path of wheat dispersion from the Fertile Crescent (yellow shade) to East Asia. Red numbers indicate the estimated date when wheat have reached to the location. The solid arrows indicate the routes of wheat dispersion that have already been identified. The dashed arrows show the potential routes of wheat in East Asia. (Modified from Gutaker and Purugganan 2023). The map was created by the same methods as Fig. 10.2

were not documented until approximately the eighth century AD, the only clues to the introduction of wheat and its use before the eighth century are archaeological remains. Excavations conducted in 2011–2012 at the Amakashi no oka Toroku site, which corresponds to the first half of the seventh century AD in Japan, identified over 190 pieces of charred wheat seeds, accounting for the largest number of plant samples (Shoda et al. 2021). The size of the wheat seeds excavated from the Amakashi no oka Toroku site was larger than those previously found at Japanese sites and similar to the size of wheat remains from sites on the Korean Peninsula. This may indicate the introduction of a new large-grain cultivar from the Korean Peninsula to Japan, which differs from wheat that was introduced in the Yayoi period (Shoda et al. 2021).

As human migration intensified during the Age of Discovery, approximately 1400–1650 AD, wheat began to traverse between countries and continents. The genomics of wheat pathogens provide insights into the migration of wheat. Sotiropoulos et al. (2022) analyzed the genomes of 172 strains of the wheat powdery mildew fungus *Blumeria graminis* f.sp. *tritici* and showed that the Japanese strain is a hybrid of the East Asian and American strains. These findings are consistent with the introduction of wheat cultivars together with pathogens from the United States for Japanese breeding. This exemplifies the effect of human beings on the migration of wheat and its pathogens.

10.6 Recent Progress and Future Perspective of Wheat Genomics

Although wheat is an essential crop with significant historical importance, only a few ancient genomic studies have been reported. This may be attributable to the complexity of the wheat genome compared with that of other crops, such as rice and maize, primarily because of their massive genome size and allohexaploid nature, with three highly conserved duplicated homoeologous genes (e.g., in A, B, and D subgenomes). Common wheat has a genome size of approximately 15 Gbp (IWGSC 2018), and even its diploid progenitors have a genome size of 5 Gbp (Luo et al. 2017; Zhao et al. 2017; Avni et al. 2022; Abrouk et al. 2023), which is larger than that of most other crops. For example, rice has a genome of approximately 390 Mbp, which is equivalent to one chromosomal arm of wheat. More than 80% of the 15 Gbp wheat genome consists of repetitive sequences (IWGSC 2018).

Owing to the advantages of next-generation sequencing and analysis methods, the chromosome-level reference genome sequence of the representative experimental accession, Chinese Spring, was released (IWGSC 2018). At the time of writing this review, we found 50 chromosomal-scale reference genome sequences from 12 species with gene annotation information (Table 10.1). These data opened the way for modern and ancient wheat genomics.

Distinguishing homoeologous genes is one of the most challenging aspects of wheat genome analysis. Recently, methods for polyploid genomics have been established that utilize these genome sequences to accurately distinguish sequencing reads corresponding to each homoeologous gene, including PolyCat (Page et al. 2013), PolyDog (Page and Udall 2015), SNiPloid (Peralta et al. 2013), HyLiTE (Duchemin et al. 2015), HANDS/HAND2 (Mithani et al. 2013; Khan et al. 2016), HomoeoRoq (Akama et al. 2014), and EAGLE-RC (Kuo et al. 2018, 2020). Benchmarking analyses showed that EAGLE-RC and GSNAP-PolyCat outperformed the other tested pipelines, and their derived expression datasets best represented expected homoeolog expression and co-expression divergence (Hu et al. 2021). Further development of these methods will facilitate progress in wheat genomics.

New genome data enhanced the understanding of the spread of wheat. Whole genome sequence data of 795 common wheat accessions indicated that wheat spread across Eurasia between 7000 and 5000 years ago (Zhao et al. 2023). Genetic analysis of Chinese wheat landraces using the Wheat 660K array suggested that Chinese wheat spread from the northwestern Caspian Sea region to South China, adapting during its agricultural trajectory to increasingly mesic and warm climatic areas and evolving from early spring to winter growth habits (Zhou et al. 2018).

Ancient DNA (aDNA) analysis of plants has been increasingly used since the end of the twentieth century. For example, maize was used for the first aDNA target sequence capture (Ávila-Arcos et al. 2011), barley seeds and their specimens were used for the first ancient RNA virus detection and the earliest ancient epigenomic sequencing, respectively (Smith et al. 2014a, b). Wheat aDNA analysis can also

Table 10.1 Available chromosome-scale genome assembly resources of wheat and its relatives

Species	Ploidy level	Genome	Resource	References
T. aestivum	6*x*	AABBDD	Chinese Spring	IWGSC (2018), Zhu et al. (2021)
				Alonge et al. (2020)
			10+ Wheat cultivars	Walkowiak et al. (2020)
			MAGIC 8+ cultivars	Walkowiak et al. (2020)
			Fielder	Sato et al. (2021)
			Sonomez	Akpinar et al. (2022)
			Kariega	Athiyannan et al. (2022)
			Renan	Aury et al. (2022)
			Attraktion	Kale et al. (2022)
			AK58	Jia et al. (2023)
T. timopheevii ssp. *armeniacum*	6*x*	$AAGGA^mA^m$	PI 94760	Grewal et al. (2024)
T. turgidum ssp. *durum*	4*x*	AABB	Svevo	Maccaferri et al. (2019)
T. turgidum ssp. *dicoccoides*	4*x*	AABB	Zavitan	Avni et al. (2017)
Triticum urartu	2*x*	AA	G1812 (PI 428198)	Ling et al. (2018)
Triticum monococcum ssp. *aegilopoides*	2*x*	A^mA^m	TA299	Ahmed et al. 2023
Triticum monococcum ssp. *monococcum*	2*x*	A^mA^m	TA10622	Ahmed et al. 2023
			PI 306540	Wang et al. (2024)
Aegilops speltoides ssp. *speltoides*	2*x*	SS	AEG-9674-1	Avni et al. (2022)
			TS01	Li et al. (2022)
Aegilops longissima	2*x*	S^lS^l	AEG-6782-2	Avni et al. (2022)
			TL05	Li et al. (2022)
Aegilops sharonensis	2*x*	$S^{sh}S^{sh}$	AS_1644	Yu et al. (2022)
			TH02	Li et al. (2022)
Aegilops bicornis	2*x*	SS	TB01	Li et al. (2022)
Aegilops searsii	2*x*	SS	Searsii	Li et al. (2022)
Aegilops tauschii	2*x*	DD	AL8/78	Luo et al. (2017)
			AL8/78	Zhao et al. (2017)
Aegilops umbellulata	2*x*	UU	TA1851	Abrouk et al. (2023)

provide detailed information on crop domestication and human migration. However, there are few examples of aDNA analysis in wheat. Genome sequencing of Egyptian Emmer wheat remains obtained from an archaeological site approximately 3000 years ago showed genetic similarity between ancient and modern Emmer landraces, indicating that the ancient Egyptian Emmer was spread from Eastern Asia (Scott et al. 2019).

Herbaria genomics is also useful for understanding recent changes in plants. In the case of potato late blight, the *Phytophthora infestans* US-1 strain was thought to be responsible for the Irish Famine. However, herbaria genomics has revealed that a closely related but unique HERB-1 strain was the causal factor in the nineteenth-century epidemic, which persisted for over 50 years (Yoshida et al. 2013). Herbaria genomics has rarely been conducted in conjunction with ancient genomics for wheat.

Asian wheat germplasms have been under-studied and under-utilized (Balfourier et al. 2019; Sansaloni et al. 2020). The introduction of the semi-dwarf allele *Rht-D1b*, originating from the Japanese wheat cultivar Norin 10, during the Green Revolution (Peng et al. 1999), is a well-known example of how East Asia has neglected genetic resources containing variation that have accumulated since the introduction of wheat. These unique characteristics and variations are crucial for understanding the history and journey of wheat from West to the Far East. Recently, wheat genomics has gained momentum by using East Asian wheat lines. For example, the representative Japanese common wheat cultivar, Norin 61 (N61), was sequenced in the 10+ wheat genome project (Walkowiak et al. 2020). This high-quality genome assembly enabled us to reveal the unique characteristics of N61, such as a tandem array of the *pathogenesis-related 13* (*PR13*) family, novel copy number variations, pseudogenization of the florigen gene *FT1*/*VRN3*, and the association of *FT-A1* alleles with spring/winter growth habits (Shimizu et al. 2021). Notably, the genomic features of N61 differed significantly from those of Chinese Spring, an experimental Chinese cultivar from East Asia. This suggests diversification during the spread of wheat from Eurasia to Japan, in addition to the influence of the Japanese breeding programs. Further analyses to better understand how wheat species spread from Central Asia to East Asia, including Japan, will be facilitated by these unique Asian genetic materials and improved ancient genomics, including DNA extraction methods from archaeological remains.

Acknowledgements The authors thank Dr. Beat Keller and Dr. Yoshihiro Matsuoka for their critical comments on the manuscript, Japan Society for the Promotion of Science (22H02316), the Swiss National Science Foundation (310030_212551), NCCR Evolving Language #51NF40_180888, Japan Science and Technology Agency (JPMJCR16O3), Japanese Ministry of Education, Culture, Sports, Science and Technology (22H05179), JSPS KAKENHI Grant Number JP22K15161, and University Research Priority Program Evolution in Action from the University of Zurich for their support.

Competing Interests The authors have no relevant financial or non-financial interests to disclose.

References

Abrouk M, Wang Y, Cavalet-Giorsa E, Troukhan M, Kravchuk M, Krattinger SG (2023) Chromosome-scale assembly of the wild wheat relative *Aegilops umbellulata*. Sci Data 10(1):739

Ahmed HI, Heuberger M, Schoen A, Koo D-H, Quiroz-Chavez J, Adhikari L, Raupp J, Cauet S, Rodde N, Cravero C, Callot C, Lazo GR, Kathiresan N, Sharma PK, Moot I, Yadav IS, Singh L, Saripalli G, Rawat N, Datla R, Athiyannan N, Ramirez-Gonzalez RH, Uauy C, Wicker T, Tiwari VK, Abrouk M, Poland J, Krattinger SG (2023) Einkorn genomics sheds light on history of the oldest domesticated wheat. Nature 620(7975):830–838

Akagi T, Jung K, Masuda K, Shimizu KK (2022) Polyploidy before and after domestication of crop species. Curr Opin Plant Biol 69:102255

Akama S, Shimizu-Inatsugi R, Shimizu KK, Sese J (2014) Genome-wide quantification of homeolog expression ratio revealed nonstochastic gene regulation in synthetic allopolyploid Arabidopsis. Nucleic Acids Res 42(6):e46

Akpinar BA, Leroy P, Watson-Haigh N, Baumann U, Barbe V, Budak H (2022) The complete genome sequence of elite bread wheat cultivar, "Sonmez". F1000Res 11:614

Alonge M, Shumate A, Puiu D, Zimin AV, Salzberg SL (2020) Chromosome-scale assembly of the bread wheat genome reveals thousands of additional gene copies. Genetics 216(2):599–608.

Arranz-Otaegui A, Colledge S, Zapata L, Teira-Mayolini LC, Ibáñez JJ (2016) Regional diversity on the timing for the initial appearance of cereal cultivation and domestication in Southwest Asia. Proc Natl Acad Sci U S A 113(49):14001–14006

Arranz-Otaegui A, Gonzalez Carretero L, Ramsey MN, Fuller DQ, Richter T (2018) Archaeobotanical evidence reveals the origins of bread 14,400 years ago in northeastern Jordan. Proc Natl Acad Sci U S A 115(31):7925–7930

Asuke S, Umehara Y, Inoue Y, Vy TTP, Iwakawa M, Matsuoka Y, Kato K, Tosa Y (2021) Origin and dynamics of *Rwt6*, a wheat gene for resistance to nonadapted pathotypes of *Pyricularia oryzae*. Phytopathology 111(11):2023–2029

Athiyannan N, Abrouk M, Boshoff WHP, Cauet S, Rodde N, Kudrna D, Mohammed N, Bettgenhaeuser J, Botha KS, Derman SS, Wing RA, Prins R, Krattinger SG (2022) Long-read genome sequencing of bread wheat facilitates disease resistance gene cloning. Nat Genet 54(3):227–231

Aury J-M, Engelen S, Istace B, Monat C, Lasserre-Zuber P, Belser C, Cruaud C, Rimbert H, Leroy P, Arribat S, Dufau I, Bellec A, Grimbichler D, Papon N, Paux E, Ranoux M, Alberti A, Wincker P, Choulet F (2022) Long-read and chromosome-scale assembly of the hexaploid wheat genome achieves high resolution for research and breeding. Gigascience 11:giac034.

Ávila-Arcos MC, Cappellini E, Romero-Navarro JA, Wales N, Moreno-Mayar JV, Rasmussen M, Fordyce SL, Montiel R, Vielle-Calzada J-P, Willerslev E, Gilbert MTP (2011) Application and comparison of large-scale solution-based DNA capture-enrichment methods on ancient DNA. Sci Rep 1(1):1–5

Avni R, Nave M, Barad O, Baruch K, Twardziok SO, Gundlach H, Hale I, Mascher M, Spannagl M, Wiebe K, Jordan KW, Golan G, Deek J, Ben-Zvi B, Ben-Zvi G, Himmelbach A, Maclachlan RP, Sharpe AG, Fritz A, Ben-David R, Budak H, Fahima T, Korol A, Faris JD, Hernandez A, Mikel MA, Levy AA, Steffenson B, Maccaferri M, Tuberosa R, Cattivelli L, Faccioli P, Ceriotti A, Kashkush K, Pourkheirandish M, Komatsuda T, Eilam T, Sela H, Sharon A, Ohad N, Chamovitz DA, Mayer KFX, Stein N, Ronen G, Peleg Z, Pozniak CJ, Akhunov ED, Distelfeld A (2017) Wild emmer genome architecture and diversity elucidate wheat evolution and domestication. Science 357:93–97

Avni R, Lux T, Minz-Dub A, Millet E, Sela H, Distelfeld A, Deek J, Yu G, Steuernagel B, Pozniak C, Ens J, Gundlach H, Mayer KFX, Himmelbach A, Stein N, Mascher M, Spannagl M, Wulff BBH, Sharon A (2022) Genome sequences of three *Aegilops* species of the section Sitopsis reveal phylogenetic relationships and provide resources for wheat improvement. Plant J 110(1):179–192

Badaeva ED, Konovalov FA, Knüpffer H, Fricano A, Ruban AS, Kehel Z, Zoshchuk SA, Surzhikov SA, Neumann K, Graner A, Hammer K, Filatenko A, Bogaard A, Jones G, Özkan H, Kilian B (2022) Genetic diversity, distribution and domestication history of the neglected GGA^tA^t genepool of wheat. Theor Appl Genet 135(3):755–776

Balfourier F, Bouchet S, Robert S, De Oliveira R, Rimbert H, Kitt J, Choulet F, International Wheat Genome Sequencing Consortium, BreedWheat Consortium, Paux E (2019) Worldwide phylogeography and history of wheat genetic diversity. Science Advances 5(5):eaav0536

Blatter RHE, Jacomet S, Schlumbaum A (2004) About the origin of European spelt (*Triticum spelta* l.): allelic differentiation of the HMW glutenin B1-1 and A1-2 subunit genes. Theor Appl Genet 108(2):360–367

Cavalet-Giorsa E, González-Muñoz A, Athiyannan N, Holden S, Salhi A, Gardener C, Quiroz-Chávez J, Rustamova SM, Elkot AF, Patpour M, Rasheed A, Mao L, Lagudah ES, Periyannan SK, Sharon A, Himmelbach A, Reif JC, Knauft M, Mascher M, Stein N, Chayut N, Ghosh S, Perovic D, Putra A, Perera AB, Hu CY, Yu G, Ahmed HI, Laquai KD, Rivera LF, Chen R, Wang Y, Gao X, Liu S, Raupp WJ, Olson EL, Lee JY, Chhuneja P, Kaur S, Zhang P, Park RF, Ding Y, Liu DC, Li W, Nasyrova FY, Dvorak J, Abbasi M, Li M, Kumar N, Meyer WB, Boshoff WHP, Steffenson BJ, Matny O, Sharma PK, Tiwari VK, Grewal S, Pozniak CJ, Chawla HS, Ens J, Dunning LT, Kolmer JA, Lazo GR, Xu SS, Gu YQ, Xu X, Uauy C, Abrouk M, Bougouffa S, Brar GS, Wulff BBH, Krattinger SG (2024) Origin and evolution of the bread wheat D genome. Nature https://doi.org/10.1038/s41586-024-07808-z

Chapman V, Miller TE, Riley R (1976) Equivalence of the A genome of bread wheat and that of *Triticum urartu*. Genet Res 27(1):69–76

Choi JY, Platts AE, Fuller DQ, Hsing Y-I, Wing RA, Purugganan MD (2017) The rice paradox: multiple origins but single domestication in Asian rice. Mol Biol Evol 34(4):969–979

Civáň P, Brown TA (2017) A novel mutation conferring the nonbrittle phenotype of cultivated barley. New Phytol 214(1):468–472

Civáň P, Ivaničová Z, Brown TA (2013) Reticulated origin of domesticated emmer wheat supports a dynamic model for the emergence of agriculture in the fertile crescent. PLoS One 8(11):e81955

Czajkowska BI, Bogaard A, Charles M, Jones G, Kohler-Schneider M, Mueller-Bieniek A, Brown TA (2020) Ancient DNA typing indicates that the "new" glume wheat of early Eurasian agriculture is a cultivated member of the *Triticum timopheevii* group. J Archaeol Sci 123:105258

Debernardi JM, Lin H, Chuck G, Faris JD, Dubcovsky J (2017) microRNA172 plays a crucial role in wheat spike morphogenesis and grain threshability. Development 144(11):1966–1975

Dorofeev VF (1969) Spontaneous hybridization in wheat populations of Transcaucasia. Euphytica 18(3):406–416

Dubcovsky J, Dvorák J (2007) Genome plasticity a key factor in the success of polyploid wheat under domestication. Science 316(5833):1862–1866

Duchemin W, Dupont P-Y, Campbell MA, Ganley ARD, Cox MP (2015) HyLiTE: accurate and flexible analysis of gene expression in hybrid and allopolyploid species. BMC Bioinformatics 16(1):8

Dvorák J (1976) The relationship between the genome of *Triticum urartu* and the A and B genomes of *Triticum aestivum*. Can J Genet Cytol 18(2):371–377

Dvorák J, Zhang HB (1990) Variation in repeated nucleotide sequences sheds light on the phylogeny of the wheat B and G genomes. Proc Natl Acad Sci U S A 87(24):9640–9644

Dvorák J, McGuire PE, Cassidy B (1988) Apparent sources of the A genomes of wheats inferred from polymorphism in abundance and restriction fragment length of repeated nucleotide sequences. Genome 30:680–689

Dvorák J, Terlizzi P, Zhang HB, Resta P (1993) The evolution of polyploid wheats: identification of the A genome donor species. Genome 36(1):21–31

Dvorák J, Deal KR, Luo M-C, You FM, von Borstel K, Dehghani H (2012) The origin of spelt and free-threshing hexaploid wheat. J Hered 103(3):426–441

FAOstat (2022) Food and Agriculture Organization of the United Nations [WWW document]. http://faostat.fao.org. Accessed 8 Jan 2024

Faris JD, Fellers JP, Brooks SA, Gill BS (2003) A bacterial artificial chromosome contig spanning the major domestication locus *Q* in wheat and identification of a candidate gene. Genetics 164(1):311–321

Feldman M (2001) Origin of cultivated wheat. In: Bonjean AP, Angus WJ (eds) The world wheat book: a history of wheat breeding. Intercept Ltd/Lavoisier Publishing, London/Paris, pp 3–56
Feldman M, Kislev ME (2007) Domestication of emmer wheat and evolution of free-threshing tetraploid wheat. ISR J Plant Sci 55(3–4):207–221
Fujita M (2013) Wheat. In: Ukai Y, Osawa R (eds) Japanese history of breeding. Yuushokan, Tokyo, pp 48–73. (in Japanese)
Fuller DQ (2007) Contrasting patterns in crop domestication and domestication rates: recent archaeobotanical insights from the old world. Ann Bot 100(5):903–924
Gao L, Zhao G, Huang D, Jia J (2017) Candidate loci involved in domestication and improvement detected by a published 90K wheat SNP array. Sci Rep 7:44530
Gaurav K, Arora S, Silva P, Sánchez-Martín J, Horsnell R, Gao L, Brar GS, Widrig V, John Raupp W, Singh N, Wu S, Kale SM, Chinoy C, Nicholson P, Quiroz-Chávez J, Simmonds J, Hayta S, Smedley MA, Harwood W, Pearce S, Gilbert D, Kangara N, Gardener C, Forner-Martínez M, Liu J, Yu G, Boden SA, Pascucci A, Ghosh S, Hafeez AN, O'Hara T, Waites J, Cheema J, Steuernagel B, Patpour M, Justesen AF, Liu S, Rudd JC, Avni R, Sharon A, Steiner B, Kirana RP, Buerstmayr H, Mehrabi AA, Nasyrova FY, Chayut N, Matny O, Steffenson BJ, Sandhu N, Chhuneja P, Lagudah E, Elkot AF, Tyrrell S, Bian X, Davey RP, Simonsen M, Schauser L, Tiwari VK, Randy Kutcher H, Hucl P, Li A, Liu D-C, Mao L, Xu S, Brown-Guedira G, Faris J, Dvorák J, Luo M-C, Krasileva K, Lux T, Artmeier S, Mayer KFX, Uauy C, Mascher M, Bentley AR, Keller B, Poland J, Wulff BBH (2022) Population genomic analysis of *Aegilops tauschii* identifies targets for bread wheat improvement. Nat Biotechnol 40:422–431
Grewal S, Yang CY, Scholefield D, Ashling S, Ghosh S, Swarbreck D, Collins J, Yao E, Sen TZ, Wilson M, Yant L, King IP, King J (2024) Chromosome-scale genome assembly of bread wheat's wild relative *Triticum timopheevii*. Sci Data 11(1). https://doi.org/10.1038/s41597-024-03260-w
Gutaker RM, Purugganan MD (2023) Adaptation and the geographic spread of crop species. Annu Rev Plant Biol 75:679–706.
Harris DR (2011) Origins of agriculture in western central Asia: an environmental-archaeological study. University of Pennsylvania Museum of Archaeology and Anthropology, Philadelphia
He F, Pasam R, Shi F, Kant S, Keeble-Gagnere G, Kay P, Forrest K, Fritz A, Hucl P, Wiebe K, Knox R, Cuthbert R, Pozniak C, Akhunova A, Morrell PL, Davies JP, Webb SR, Spangenberg G, Hayes B, Daetwyler H, Tibbits J, Hayden M, Akhunov E (2019) Exome sequencing highlights the role of wild-relative introgression in shaping the adaptive landscape of the wheat genome. Nat Genet 51(5):896–904
Heun M, Schäfer-Pregl R, Klawan D, Castagna R, Accerbi M, Borghi B, Salamini F (1997) Site of einkorn wheat domestication identified by DNA fingerprinting. Science 278(5341):1312–1314
Hirosawa S, Takumi S, Ishii T, Kawahara T, Nakamura C, Mori N (2004) Chloroplast and nuclear DNA variation in common wheat: insight into the origin and evolution of common wheat. Genes Genet Syst 79(5):271–282
Hu G, Grover CE, Arick MA, Liu M, Peterson DG, Wendel JF (2021) Homoeologous gene expression and co-expression network analyses and evolutionary inference in allopolyploids. Brief Bioinform 22(2):1819-1835
Huang S, Sirikhachornkit A, Su X, Faris J, Gill B, Haselkorn R, Gornicki P (2002) Genes encoding plastid acetyl-CoA carboxylase and 3-phosphoglycerate kinase of the *Triticum/Aegilops* complex and the evolutionary history of polyploid wheat. Proc Natl Acad Sci U S A 99(12):8133–8138
Jia J, Zhao G, Li D, Wang K, Kong C, Deng P, Yan X, Zhang X, Lu Z, Xu S, Jiao Y, Chong K, Liu X, Cui D, Li G, Zhang Y, Du C, Wu L, Li T, Yan D, Zhan K, Chen F, Wang Z, Zhang L, Kong X, Ru Z, Wang D, Gao L (2023) Genome resources for the elite bread wheat cultivar Aikang 58 and mining of elite homeologous haplotypes for accelerating wheat improvement. Mol Plant 16(12):1893–1910
Kale SM, Schulthess AW, Padmarasu S, Boeven PHG, Schacht J, Himmelbach A, Steuernagel B, Wulff BBH, Reif JC, Stein N, Mascher M (2022) A catalogue of resistance gene homologs

and a chromosome-scale reference sequence support resistance gene mapping in winter wheat. Plant Biotechnol J 20(9):1730–1742

Khan A, Belfield EJ, Harberd NP, Mithani A (2016) HANDS2: accurate assignment of homoeallelic base-identity in allopolyploids despite missing data. Sci Rep 6:29234

Kihara H (1944) Discovery of the DD-analyser, one of the ancestors of *Triticum vulgare*. Agric Hortic 19:889

Kihara H (1966) Factors affecting the evolution of common wheat. Indian J Genet Plant Breed 26A:14–28

Kihara H, Lilienfeld F (1934) Genomanalyse bei *Triticum* und *Aegilops* v. *Triticum timopheevi* Zhuk. Cytologia 6(1):87–122

Kilian B, Ozkan H, Deusch O, Effgen S, Brandolini A, Kohl J, Martin W, Salamini F (2007a) Independent wheat B and G genome origins in outcrossing *Aegilops* progenitor haplotypes. Mol Biol Evol 24(1):217–227

Kilian B, Ozkan H, Walther A, Kohl J, Dagan T, Salamini F, Martin W (2007b) Molecular diversity at 18 loci in 321 wild and 92 domesticate lines reveal no reduction of nucleotide diversity during *Triticum monococcum* (einkorn) domestication: implications for the origin of agriculture. Mol Biol Evol 24(12):2657–2668

Kuckuck H (1959) On the findings of *Triticum spelta* l. in Iran and on the arising of *Triticum aestivum*-types through crossing of different spelta-types. Wheat Inf Serv 9–10:1–2

Kuijt I, Finlayson B (2009) Evidence for food storage and predomestication granaries 11,000 years ago in the Jordan valley. Proc Natl Acad Sci U S A 106(27):10966–10970

Kuo T, Frith MC, Sese J, Horton P (2018) EAGLE: explicit alternative genome likelihood evaluator. BMC Med Genet 11(Suppl 2):28

Kuo TCY, Hatakeyama M, Tameshige T, Shimizu KK, Sese J (2020) Homeolog expression quantification methods for allopolyploids. Brief Bioinform 21(2):395–407

Levy AA, Feldman M (2022) Evolution and origin of bread wheat. Plant Cell 34(7):2549–2567

Lev-Yadun S, Gopher A, Abbo S (2000) The cradle of agriculture. Science 288(5471):1602–1603

Li L-F, Zhang Z-B, Wang Z-H, Li N, Sha Y, Wang X-F, Ding N, Li Y, Zhao J, Wu Y, Gong L, Mafessoni F, Levy AA, Liu B (2022) Genome sequences of five Sitopsis species of *Aegilops* and the origin of polyploid wheat B subgenome. Mol Plant 15(3):488–503

Ling HQ, Ma B, Shi X, Liu H, Dong L, Sun H, Cao Y, Gao Q, Zheng S, Li Y, Yu Y, Du H, Qi M, Li Y, Lu H, Yu H, Cui Y, Wang N, Chen C, Wu H, Zhao Y, Zhang J, Li Y, Zhou W, Zhang B, Hu W, van Eijk MJT, Tang J, Witsenboer HMA, Zhao S, Li Z, Zhang A, Wang D, Liang C (2018) Genome sequence of the progenitor of wheat A subgenome *Triticum urartu*. Nature 557(7705):424–428

Liu X, Lister DL, Zhao Z, Petrie CA, Zeng X, Jones PJ, Staff RA, Pokharia AK, Bates J, Singh RN, Weber SA, Motuzaite Matuzeviciute G, Dong G, Li H, Lü H, Jiang H, Wang J, Ma J, Tian D, Jin G, Zhou L, Wu X, Jones MK (2017) Journey to the east: diverse routes and variable flowering times for wheat and barley en route to prehistoric China. PLoS One 12(11):e0187405

Luo M-C, Yang Z-L, You FM, Kawahara T, Waines JG, Dvorák J (2007) The structure of wild and domesticated emmer wheat populations, gene flow between them, and the site of emmer domestication. Theor Appl Genet 114(6):947–959

Luo M-C, Gu YQ, Puiu D, Wang H, Twardziok SO, Deal KR, Huo N, Zhu T, Wang L, Wang Y, McGuire PE, Liu S, Long H, Ramasamy RK, Rodriguez JC, Van SL, Yuan L, Wang Z, Xia Z, Xiao L, Anderson OD, Ouyang S, Liang Y, Zimin AV, Pertea G, Qi P, Bennetzen JL, Dai X, Dawson MW, Müller H-G, Kugler K, Rivarola-Duarte L, Spannagl M, Mayer KFX, Lu F-H, Bevan MW, Leroy P, Li P, You FM, Sun Q, Liu Z, Lyons E, Wicker T, Salzberg SL, Devos KM, Dvořák J (2017) Genome sequence of the progenitor of the wheat D genome *Aegilops tauschii*. Nature 551(7681):498–502

Maccaferri M, Harris NS, Twardziok SO, Pasam RK, Gundlach H, Spannagl M, Ormanbekova D, Lux T, Prade VM, Milner SG, Himmelbach A, Mascher M, Bagnaresi P, Faccioli P, Cozzi P, Lauria M, Lazzari B, Stella A, Manconi A, Gnocchi M, Moscatelli M, Avni R, Deek J, Biyiklioglu S, Frascaroli E, Corneti S, Salvi S, Sonnante G, Desiderio F, Marè C, Crosatti C, Mica E, Özkan H, Kilian B, De Vita P, Marone D, Joukhadar R, Mazzucotelli E, Nigro D, Gadaleta A, Chao S, Faris JD, Melo ATO, Pumphrey M, Pecchioni N, Milanesi L, Wiebe K, Ens

J, MacLachlan RP, Clarke JM, Sharpe AG, Koh CS, Liang KYH, Taylor GJ, Knox R, Budak H, Mastrangelo AM, Xu SS, Stein N, Hale I, Distelfeld A, Hayden MJ, Tuberosa R, Walkowiak S, Mayer KFX, Ceriotti A, Pozniak CJ, Cattivelli L (2019) Durum wheat genome highlights past domestication signatures and future improvement targets. Nat Genet 51(5):885–895

MacKey J (1966) Species relationship in Triticum. In: MacKey J (ed) Proceedings of the 2nd international wheat genetics symposium. Mendelian Society of Lund for Scandinavian Association of Geneticists, Lund, pp 237–275

Mastrangelo AM, Cattivelli L (2021) What makes bread and durum wheat different? Trends Plant Sci 26(7):677–684

Matsuoka Y (2011) Evolution of polyploid *Triticum* wheats under cultivation: the role of domestication, natural hybridization and allopolyploid speciation in their diversification. Plant Cell Physiol 52:750–764

Matsuoka Y, Jaffar Aghaei M, Abbasi MR, Totiaei A, Mozafari J, Ohta S (2008) Durum wheat cultivation associated with *Aegilops tauschii* in northern Iran. Genet Resour Crop Evol 55(6):861–868

McFadden ES, Sears ER (1944) The artificial synthesis of *Triticum spelta*. Rec Genet Soc Am 13:26–27

Mithani A, Belfield EJ, Brown C, Jiang C, Leach LJ, Harberd NP (2013) HANDS: a tool for genome-wide discovery of subgenome-specific base-identity in polyploids. BMC Genomics 14:653

Miyashita NT, Mori N, Tsunewaki K (1994) Molecular variation in chloroplast DNA regions in ancestral species of wheat. Genetics 137(3):883–889

Molina J, Sikora M, Garud N, Flowers JM, Rubinstein S, Reynolds A, Huang P, Jackson S, Schaal BA, Bustamante CD, Boyko AR, Purugganan MD (2011) Molecular evidence for a single evolutionary origin of domesticated rice. Proc Natl Acad Sci U S A 108(20):8351–8356

Mori N, Ishii T, Ishido T, Hirosawa S, Watatani H, Kawahara T, Nesbitt M, Belay G, Takumi S, Ogihara Y (2003) Origin of domesticated emmer and common wheat inferred from chloroplast DNA fingerprinting. Proc Xth Int Wheat Genet Symp 1:25–28

Mori N, Kondo Y, Ishii T, Kawahara T, Valkoun J, Nakamura C (2009) Genetic diversity and origin of *Timopheevi* wheat inferred by chloroplast DNA fingerprinting. Breed Sci 59(5):571–578

Muramatsu M (1979) Presence of vulgare gene, *Q*, in a dense-spike variety of *Triticum dicoccum* Schubl. Rep Plant Germplasm Inst Kyoto Univ 4:39–41

Muramatsu M (1985) Spike type in two cultivars of *Triticum dicoccum* with the spelta gene *Q* compared with the Q-bearing variety liguliforme. Jpn J Breed 35(3):255–267

Muramatsu M (1986) The vulgare super gene, *Q*: its universality in durum wheat and its phenotypic effects in tetraploid and hexaploid wheats. Can J Genet Cytol 28(1):30–41

Nave M, Avni R, Ben-Zvi B, Hale I, Distelfeld A (2016) QTLs for uniform grain dimensions and germination selected during wheat domestication are co-located on chromosome 4B. Theor Appl Genet 129(7):1303–1315

Nave M, Taş M, Raupp J, Tiwari VK, Ozkan H, Poland J, Hale I, Komatsuda T, Distelfeld A (2021) The independent domestication of Timopheev's wheat: insights from haplotype analysis of the *brittle rachis 1* (*BTR1-A*) gene. Genes (Basel) 12(3):338

Nesbitt M (2001) Wheat evolution: integrating archaeological and biological evidence. In: Caligari PDS, Brandham PE (eds) Wheat taxonomy: the legacy of John Percival. Linnean Society, London, pp 37–59

Nishimura K, Motoki K, Yamazaki A, Takisawa R, Yasui Y, Kawai T, Ushijima K, Nakano R, Nakazaki T (2022) MIG-seq is an effective method for high-throughput genotyping in wheat (*Triticum* spp.). DNA Res 29(2):dsac011.

Noda K, Ge K (1989) Structural differentiation of chromosomes between *Triticum dicoccoides* Koern. and *T. araraticum* Jakubz., showing high meiotic pairing homology. Jpn J Genet 64(3):191–198

Ogihara Y, Tsunewaki K (1988) Diversity and evolution of chloroplast DNA in *Triticum* and *Aegilops* as revealed by restriction fragment analysis. Theor Appl Genet 76(3):321–332

Ozkan H, Brandolini A, Schäfer-Pregl R, Salamini F (2002) AFLP analysis of a collection of tetraploid wheats indicates the origin of emmer and hard wheat domestication in southeast Turkey. Mol Biol Evol 19(10):1797–1801
Ozkan H, Brandolini A, Pozzi C, Effgen S, Wunder J, Salamini F (2005) A reconsideration of the domestication geography of tetraploid wheats. Theor Appl Genet 110(6):1052–1060
Page JT, Udall JA (2015) Methods for mapping and categorization of DNA sequence reads from allopolyploid organisms. BMC Genet 16(Suppl 2):S4
Page JT, Gingle AR, Udall JA (2013) PolyCat: a resource for genome categorization of sequencing reads from allopolyploid organisms. G3 3(3):517–525
Peng J, Richards DE, Hartley NM, Murphy GP, Devos KM, Flintham JE, Beales J, Fish LJ, Worland AJ, Pelica F, Sudhakar D, Christou P, Snape JW, Gale MD, Harberd NP (1999) 'Green revolution' genes encode mutant gibberellin response modulators. Nature 400(6741):256–261
Peralta M, Combes M-C, Cenci A, Lashermes P, Dereeper A (2013) SNiPloid: a utility to exploit high-throughput SNP data derived from RNA-Seq in allopolyploid species. Int J Plant Genomics 2013:890123
Pont C, Leroy T, Seidel M, Tondelli A, Duchemin W, Armisen D, Lang D, Bustos-Korts D, Goué N, Balfourier F, Molnár-Láng M, Lage J, Kilian B, Özkan H, Waite D, Dyer S, Letellier T, Alaux M, Wheat and Barley Legacy for Breeding Improvement (WHEALBI) consortium, Russell J, Keller B, van Eeuwijk F, Spannagl M, Mayer KFX, Waugh R, Stein N, Cattivelli L, Haberer G, Charmet G, Salse J (2019) Tracing the ancestry of modern bread wheats. Nat Genet 51(5):905–911
Pourkheirandish M, Komatsuda T (2007) The importance of barley genetics and domestication in a global perspective. Ann Bot 100(5):999–1008
Pourkheirandish M, Hensel G, Kilian B, Senthil N, Chen G, Sameri M, Azhaguvel P, Sakuma S, Dhanagond S, Sharma R, Mascher M, Himmelbach A, Gottwald S, Nair SK, Tagiri A, Yukuhiro F, Nagamura Y, Kanamori H, Matsumoto T, Willcox G, Middleton CP, Wicker T, Walther A, Waugh R, Fincher GB, Stein N, Kumlehn J, Sato K, Komatsuda T (2015) Evolution of the grain dispersal system in barley. Cell 162(3):527–539
Pourkheirandish M, Dai F, Sakuma S, Kanamori H, Distelfeld A, Willcox G, Kawahara T, Matsumoto T, Kilian B, Komatsuda T (2017) On the origin of the non-brittle rachis trait of domesticated einkorn wheat. Front Plant Sci 8:2031
Riehl S, Zeidi M, Conard NJ (2013) Emergence of agriculture in the foothills of the Zagros mountains of Iran. Science 341(6141):65–67
Sakuma S, Salomon B, Komatsuda T (2011) The domestication syndrome genes responsible for the major changes in plant form in the Triticeae crops. Plant Cell Physiol 52(5):738–749
Salamini F, Özkan H, Brandolini A, Schäfer-Pregl R, Martin W (2002) Genetics and geography of wild cereal domestication in the near east. Nat Rev Genet 3(6):429–441
Sansaloni C, Franco J, Santos B, Percival-Alwyn L, Singh S, Petroli C, Campos J, Dreher K, Payne T, Marshall D, Kilian B, Milne I, Raubach S, Shaw P, Stephen G, Carling J, Pierre CS, Burgueño J, Crosa J, Li H, Guzman C, Kehel Z, Amri A, Kilian A, Wenzl P, Uauy C, Banziger M, Caccamo M, Pixley K (2020) Diversity analysis of 80,000 wheat accessions reveals consequences and opportunities of selection footprints. Nat Commun 11(1):4572
Sarkar P, Stebbins GL (1956) Morphological evidence concerning the origin of the B genome in wheat. Am J Bot 43(4):297–304
Sato K, Abe F, Mascher M, Haberer G, Gundlach H, Spannagl M, Shirasawa K, Isobe S (2021) Chromosome-scale genome assembly of the transformation-amenable common wheat cultivar 'Fielder'. DNA Res 28(3):dsab008
Scott MF, Botigué LR, Brace S, Stevens CJ, Mullin VE, Stevenson A, Thomas MG, Fuller DQ, Mott R (2019) A 3,000-year-old Egyptian emmer wheat genome reveals dispersal and domestication history. Nat Plants 5(11):1120–1128
Sears ER (1954) The aneuploids of common wheat. Missouri Agric Exp Stn Res Bull 572:1–58
Shewry PR, Hey SJ (2015) The contribution of wheat to human diet and health. Food Energy Secur 4(3):178–202

Shimizu KK (2022) Robustness and the generalist niche of polyploid species: genome shock or gradual evolution? Curr Opin Plant Biol 69:102292

Shimizu KK, Copetti D, Okada M, Wicker T, Tameshige T, Hatakeyama M, Shimizu-Inatsugi R, Aquino C, Nishimura K, Kobayashi F, Murata K, Kuo T, Delorean E, Poland J, Haberer G, Spannagl M, Mayer KFX, Gutierrez-Gonzalez J, Muehlbauer GJ, Monat C, Himmelbach A, Padmarasu S, Mascher M, Walkowiak S, Nakazaki T, Ban T, Kawaura K, Tsuji H, Pozniak C, Stein N, Sese J, Nasuda S, Handa H (2021) *De novo* genome assembly of the Japanese wheat cultivar Norin 61 highlights functional variation in flowering time and *Fusarium*-resistant genes in East Asian genotypes. Plant Cell Physiol 62(1):8–27

Shoda S, Bhandari S, Sasaki Y, Murakami N, Liu X (2021) Archaeobotanical study on the charred wheat grains from the Amakashi no oka Toroku site, Japan: a Eurasian perspective. Papers Nara Natl Res Inst Cult Prop 2:29–65

Simons KJ, Fellers JP, Trick HN, Zhang Z, Tai Y-S, Gill BS, Faris JD (2006) Molecular characterization of the major wheat domestication gene *Q*. Genetics 172(1):547–555

Smith BD (1998) The emergence of agriculture. Scientific American Library, New York, pp 1–13

Smith O, Clapham AJ, Rose P, Liu Y, Wang J, Allaby RG (2014a) Genomic methylation patterns in archaeological barley show de-methylation as a time-dependent diagenetic process. Sci Rep 4(1):1–6

Smith O, Clapham A, Rose P, Liu Y, Wang J, Allaby RG (2014b) A complete ancient RNA genome: identification, reconstruction and evolutionary history of archaeological barley stripe mosaic virus. Sci Rep 4(1):1–6

Sotiropoulos AG, Arango-Isaza E, Ban T, Barbieri C, Bourras S, Cowger C, Czembor PC, Ben-David R, Dinoor A, Ellwood SR, Graf J, Hatta K, Helguera M, Sánchez-Martín J, McDonald BA, Morgounov AI, Müller MC, Shamanin V, Shimizu KK, Yoshihira T, Zbinden H, Keller B, Wicker T (2022) Global genomic analyses of wheat powdery mildew reveal association of pathogen spread with historical human migration and trade. Nat Commun 13(1):4315

Stoneking M (2017) An introduction to molecular anthropology. Wiley-Blackwell, Hoboken

Sun J, Okada M, Tameshige T, Shimizu-Inatsugi R, Akiyama R, Nagano AJ, Sese J, Shimizu KK (2023) A low-coverage 3' RNA-seq to detect homeolog expression in polyploid wheat. NAR Genom Bioinform 5(3):lqad067

Svetozarova VV (1939) Second genome of *Triticum timopheevi* zhuk. C R Acad Sci USSR 23:473–477

The International Wheat Genome Sequence Consortium (IWGSC) (2018) Shifting the limits in wheat research and breeding using a fully annotated reference genome. Science 361(6403):eaar7191

Tsunewaki K (1968) Origin and phylogenetic differentiation of common wheat revealed by comparative gene analysis. In: Proc 3rd Int Wheat Genet Symp, Canberra, pp 71–85

Walkowiak S, Gao L, Monat C, Haberer G, Kassa MT, Brinton J, Ramirez-Gonzalez RH, Kolodziej MC, Delorean E, Thambugala D, Klymiuk V, Byrns B, Gundlach H, Bandi V, Siri JN, Nilsen K, Aquino C, Himmelbach A, Copetti D, Ban T, Venturini L, Bevan M, Clavijo B, Koo D-H, Ens J, Wiebe K, N'Diaye A, Fritz AK, Gutwin C, Fiebig A, Fosker C, Fu BX, Accinelli GG, Gardner KA, Fradgley N, Gutierrez-Gonzalez J, Halstead-Nussloch G, Hatakeyama M, Koh CS, Deek J, Costamagna AC, Fobert P, Heavens D, Kanamori H, Kawaura K, Kobayashi F, Krasileva K, Kuo T, McKenzie N, Murata K, Nabeka Y, Paape T, Padmarasu S, Percival-Alwyn L, Kagale S, Scholz U, Sese J, Juliana P, Singh R, Shimizu-Inatsugi R, Swarbreck D, Cockram J, Budak H, Tameshige T, Tanaka T, Tsuji H, Wright J, Wu J, Steuernagel B, Small I, Cloutier S, Keeble-Gagnère G, Muehlbauer G, Tibbets J, Nasuda S, Melonek J, Hucl PJ, Sharpe AG, Clark M, Legg E, Bharti A, Langridge P, Hall A, Uauy C, Mascher M, Krattinger SG, Handa H, Shimizu KK, Distelfeld A, Chalmers K, Keller B, Mayer KFX, Poland J, Stein N, McCartney CA, Spannagl M, Wicker T, Pozniak CJ (2020) Multiple wheat genomes reveal global variation in modern breeding. Nature 588(7837):277–283

Wang X, Li H, Shen T, Wang X, Yi S, Meng T, Sun J, Wang X, Qu X, Chen S, Guo L (2024) A near-complete genome sequence of einkorn wheat provides insight into the evolution of wheat A subgenomes. Plant Commun 5(5):100768

Willcox G (2012) Pre-domestic cultivation during the Late Pleistocene and Early Holocene in the Northern Levant. In: Gepts P, Famula TR, Bettinger RL et al (eds) Biodiversity in agriculture: domestication, evolution, and sustainability. Cambridge University Press, Cambridge, pp 92–109

Yadav IS, Singh N, Wu S, Raupp J, Wilson DL, Rawat N, Gill BS, Poland J, Tiwari VK (2023) Exploring genetic diversity of wild and related tetraploid wheat species *Triticum turgidum* and *Triticum timopheevii*. J Advert Res 48:47–60

Yoshida K, Schuenemann VJ, Cano LM, Pais M, Mishra B, Sharma R, Lanz C, Martin FN, Kamoun S, Krause J, Thines M, Weigel D, Burbano HA (2013) The rise and fall of the *Phytophthora infestans* lineage that triggered the Irish potato famine. Elife 2:e00731

Yu G, Matny O, Champouret N, Steuernagel B, Moscou MJ, Hernández-Pinzón I, Green P, Hayta S, Smedley M, Harwood W, Kangara N, Yue Y, Gardener C, Banfield MJ, Olivera PD, Welchin C, Simmons J, Millet E, Minz-Dub A, Ronen M, Avni R, Sharon A, Patpour M, Justesen AF, Jayakodi M, Himmelbach A, Stein N, Wu S, Poland J, Ens J, Pozniak C, Karafiátová M, Molnár I, Doležel J, Ward ER, Reuber TL, Jones JDG, Mascher M, Steffenson BJ, Wulff BBH (2022) *Aegilops sharonensis* genome-assisted identification of stem rust resistance gene *Sr62*. Nat Commun 13(1). https://doi.org/10.1038/s41467-022-29132-8

Zhao G, Zou C, Li K, Wang K, Li T, Gao L, Zhang X, Wang H, Yang Z, Liu X, Jiang W, Mao L, Kong X, Jiao Y, Jia J (2017) The *Aegilops tauschii* genome reveals multiple impacts of transposons. Nat Plants 3(12):946–955

Zhao Y, Xie P, Guan P, Wang Y, Li Y, Yu K, Xin M, Hu Z, Yao Y, Ni Z, Sun Q, Xie C, Peng H (2019) *Btr1-A* induces grain shattering and affects spike morphology and yield-related traits in wheat. Plant Cell Physiol 60(6):1342–1353

Zhao X, Guo Y, Kang L, Yin C, Bi A, Xu D, Zhang Z, Zhang J, Yang X, Xu J, Xu S, Song X, Zhang M, Li Y, Kear P, Wang J, Liu Z, Fu X, Lu F (2023) Population genomics unravels the holocene history of bread wheat and its relatives. Nat Plants 9(3):403–419

Zhou Y, Chen Z, Cheng M, Chen J, Zhu T, Wang R, Liu Y, Qi P, Chen G, Jiang Q, Wei Y, Luo M-C, Nevo E, Allaby RG, Liu D, Wang J, Dvorák J, Zheng Y (2018) Uncovering the dispersion history, adaptive evolution and selection of wheat in China. Plant Biotechnol J 16(1):280–291

Zhou X, Yu J, Spengler RN, Shen H, Zhao K, Ge J, Bao Y, Liu J, Yang Q, Chen G, Weiming Jia P, Li X (2020a) 5,200-year-old cereal grains from the eastern Altai mountains redate the trans-Eurasian crop exchange. Nat Plants 6(2):78–87

Zhou Y, Zhao X, Li Y, Xu J, Bi A, Kang L, Xu D, Chen H, Wang Y, Wang Y-G, Liu S, Jiao C, Lu H, Wang J, Yin C, Jiao Y, Lu F (2020b) *Triticum* population sequencing provides insights into wheat adaptation. Nat Genet 52(12):1412–1422

Zhu T, Wang L, Rimbert H, Rodriguez JC, Deal KR, De Oliveira R, Choulet F, Keeble-Gagnère G, Tibbits J, Rogers J, Eversole K, Appels R, Gu YQ, Mascher M, Dvorák J, Luo M-C (2021) Optical maps refine the bread wheat *Triticum aestivum* cv. Chinese Spring genome assembly. Plant J 107(1):303–314

Zimin AV, Puiu D, Hall R, Kingan S, Clavijo BJ, Salzberg SL (2017) The first near-complete assembly of the hexaploid bread wheat genome, *Triticum aestivum*. Gigascience 6(11):1–7

Zohary D, Hopf M (1993) Domestication of plants in the old world. Clarendon Press, Oxford

Zohary D, Hopf M (2000) Domestication of plants in the old world, 3rd edn. Oxford University Press, New York, pp 42–51

Chapter 11
Geographical Distribution of Animal and Crop Terms in Asian and African Languages

Mitsuaki Endo, Hiroyuki Suzuki, and Chitsuko Fukushima

Abstract This chapter discusses the geographical distribution of 13 word items related to domesticated animals and crops in Asia and Africa with geolinguistic method: (1) rat/mouse, (2) chicken, (3) horse, (4) dog, (5) wolf, (6) bear, (7) rice plant, (8) wheat, (9) broomcorn millet, (10) foxtail millet, (11) barnyard millet, (12) taro, and (13) yam. Individual maps show representative word forms from each language family. Previous interpretations of their interrelationships are summarized, along with the authors' opinions, if available. This study focuses on the relationships between the Japonic language family and its surrounding language families in East Asia. In conclusion, the absence of these crops and words in the outer areas clearly defined the Japonic-speaking people's homeland.

Keywords Wanderwort · Geolinguistics · Borrowing · Language families · Language universals

---

M. Endo (✉)
Aoyama Gakuin University, Shibuya-ku, Tokyo, Japan
e-mail: t10976@aoyamagakuin.jp

H. Suzuki
Kyoto University, Kyoto, Japan

C. Fukushima
University of Niigata Prefecture, Niigata, Japan

N. Osada et al. (eds.), *Phylogeographic History of Plants and Animals Coexisting with Humans in Asia*, Evolutionary Studies,
https://doi.org/10.1007/978-981-97-6887-5_11

Abbreviations

LAA	Mitsuaki Endo, Makoto Minegishi, Satoko Shirai, Hiroyuki Suzuki, and Keita Kurabe eds. 2021. *Linguistic Atlas of Asia*. Tokyo: Hituzi Syobo Publishing
LAAA-1	Hiroyuki Suzuki, Mika Fukazawa, Akiko Yokoyama, and Mitsuaki Endo, eds. 2022. *Linguistic Atlas of Asia and Africa*, Vol. 1. Tokyo: Geolinguistic Society of Japan. https://doi.org/10.5281/zenodo.7118188
LAAA-2	Hiroyuki Suzuki, Kohei Nakazawa, and Mitsuaki Endo, eds. 2023. *Linguistic Atlas of Asia and Africa*, Vol. 2. Tokyo: Geolinguistic Society of Japan. https://doi.org/10.5281/zenodo.7754469
STEDT	Sino-Tibetan Etymological Dictionary and Thesaurus. https://stedt.berkeley.edu/

11.1 Introduction

This chapter uses a geolinguistic approach to examine animal and crop terms in Asia and Africa. Geolinguistics is a recent term for dialect geography or linguistic geography that focuses on multilingual treatment. Similar to phylogeography in biology, it aims to identify geographical distributions of words or other linguistic features, of which we inquire into the formation process. We examine each map individually and then extract the main linguistic adaptation methods to encounter new terms brought about by domesticating wild animals and crops with special reference to the development of the Japanese language (or Japonic).

The main clue for detecting the diffusion of particular words is the so-called "Wanderwort," i.e., words borrowed sporadically beyond languages or language families. Well-known examples are "tea" and "coca-cola," which spread worldwide. In the case of "coca-cola," the resemblance of word forms is clear, so there is no need for scrutiny. As for ancient loanwords, a series of sound changes hid the original forms; hence, it is necessary to reconstruct the old forms by applying regular sound correspondence rules among related languages in comparative linguistics. Moreover, word forms tend to be deformed when borrowed due to differences in the donor and acceptor phonological systems.

Thirteen items will be examined: rat/mouse, chicken, horse, dog, wolf, bear, rice plant, wheat, broomcorn millet, foxtail millet, barnyard millet, taro, and yam. These words have already been mapped and explained in the *Linguistic Atlas of Asia* (hereafter abbreviated as LAA) and the *Linguistic Atlas of Asia and Africa*, Volumes 1 and 2 (LAAA-1, -2). The original maps were created for each language or language family with minute symbols in approximately 2000 localities. This chapter summarizes the general macrotendency of the geographical distributions of these 13 words, providing an overview of the interpretations of the interrelationships among these words.

11.2 Methods

The geographical distributions of 13 items introduced above are shown in maps with 22 representative points, as shown in Fig. 11.1. Most of these names are language families, although some are areas. For example, "South Asia" includes Indo-Aryan, Nuristani, Andamanese, and language isolates such as Burushaski, Kusunda, Nihali, and Vedda; "Kalahari Basin Area" includes three language families, Tuu, Kx'a, and Khoe Kwadi; and "Caucasian" comprises distinct language families of Kartvelian, Abkhazo-Adyghean, and Nakho-Daghestanian. In other cases, language family branches are divided into two areas: Indo-European into South Asia and Iranian and Sino-Tibetan into Sinitic and Tibeto-Burman. Such groupings and names were adopted based on the judgments of the linguists in charge of each group. Since these correspond to the chapters in LAA, LAAA-1, and LAAA-2, it is convenient to consult the micro-geographical distributions in these volumes to determine if more precise information is needed.

For simplicity, English words are used as the index of each item. Specifically, the real unit is "a semantic field," not an English word. For example, "rice" in English possibly means "rice plant" (*ine* in Japanese), "rice without husk" (*kome* in Japanese), and "rice with husk" (*momi* in Japanese), while these three words are distinguished with totally different meanings in some languages, including Japanese. It is possible to establish a semantic field using English for "rice" and further dividing it into microsemantic fields such as "rice plant," "rice without husk," and "rice

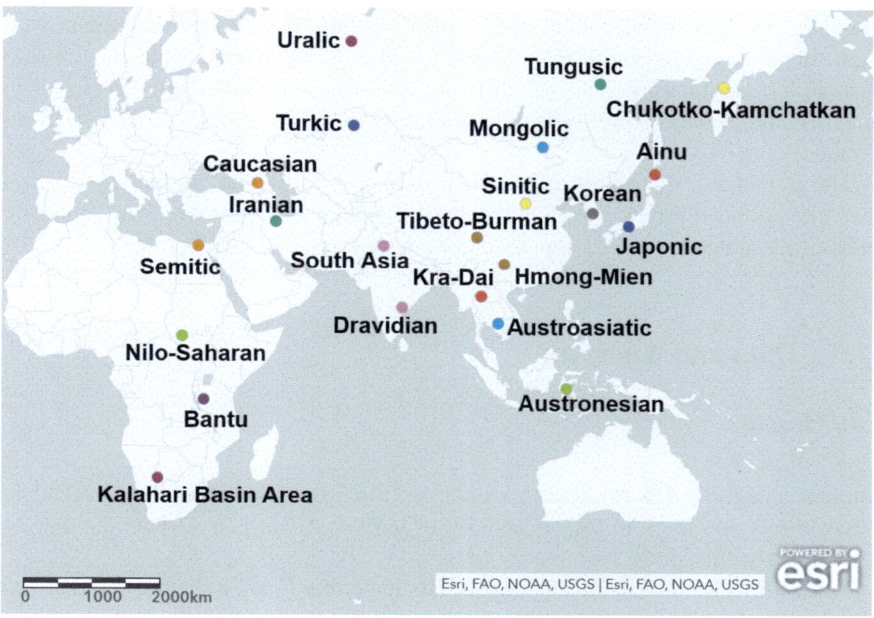

Fig. 11.1 Location of language families and language areas

with husk." Another case is that many languages do not distinguish between mouse and rat as in English. How to divide a semantic field differs depending on the language or dialect. We encounter situations where it is difficult to judge which word form covers which semantic field.

The criterion for determining the identity between the word forms is the resemblance of their sound shape. Although this is a vague definition, it is difficult to determine how much resemblance is enough to eliminate the possibility of coincidence. Suppose that the average number of consonants is 10 and of vowels 5 in a particular language; in that case, the chance of resemblance of a consonant is 1/10, a vowel 1/5, a consonant plus a vowel 1/50, two consonants 1/100, two consonants plus a vowel 1/500, and so on. Although *namae,* "name," in Japanese resembles two consonants plus two vowels with *name* in English, it is considered a typical example of coincidence between the two languages. Yet *ga,* "railway station," in Vietnamese consists of a consonant plus a vowel only; nevertheless, it is undoubtedly a loanword from French *gare,* "railway station," since the French constructed the railways, and there are other French loanwords in the Vietnamese language. If the part of the resemblance is short, it is necessary to add other circumstantial evidence. Direct contact between two languages at the same place and time is also required. However, migration or extinction of the intervening languages occasionally hide such relationships, and only hypotheses can be raised, despite missing links.

If there were cognate words within a language family or systematic loanwords such as Latin words in English or Chinese words in Japanese, abundant parallel examples can be found for each sound consisting of a word; therefore, sound laws can be prescribed. As for the case of sporadic loanwords, it is rare to find parallel examples. In practice, we collect and compare previous theories on individual cases. If an unknown resemblance is detected, we also mention a possible relationship through a heuristic approach, which is not decisive. Additionally, recent English, Spanish, and Russian loanwords are treated as "new loanwords," differently from "Wanderwort."

The homeland, dispersion routes, and dating of individual animals and crops have already been inferred from archeological and genetic evidence. The linguistic paths of these items that accompanied human groups are shown below.

11.3 Data and Analyses

Rat/Mouse

Animals categorized as *Rodentia* are diverse. However, it is difficult to differentiate between rats and mice in the vocabulary of various Asian and African languages; consequently, this study examines both types.

Word forms used by each language or language family are indicated on the location. This is the method used by the *Atlas linguistique de la France*. In many cases, the index words represent many other phonetic variants. Loanwords, semantic

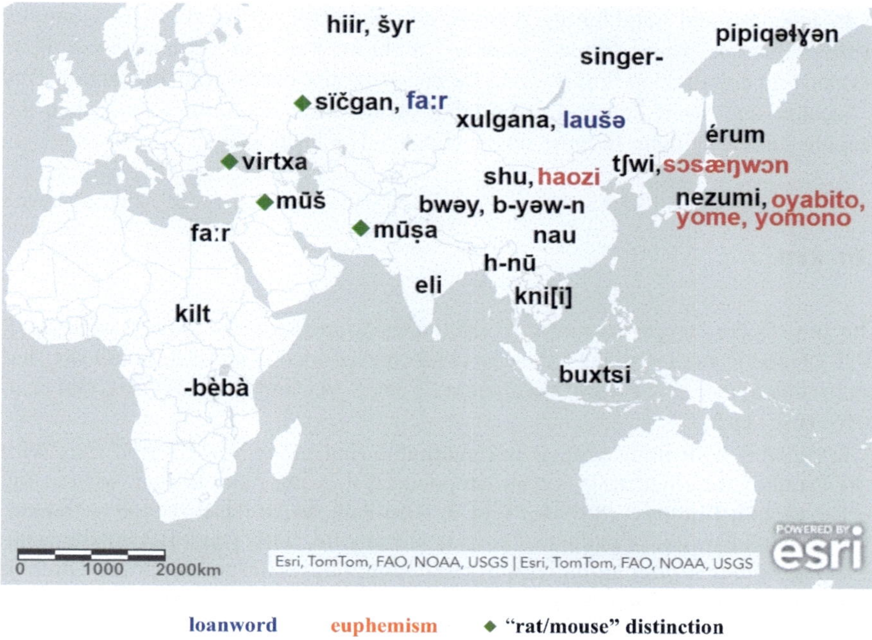

Fig. 11.2 Rat/mouse

transfers, semantic distinctions, euphemisms, or honorifics are indicated by colored letters or symbols.

Many word forms appear in Fig. 11.2. This is an ordinary case as a linguistic map because different words emerge independently inside a language family by nature.

Three word-types start from *b*; however, the rest of these word forms have no similarity and are scattered throughout isolated places.

Schuessler (2007: 471–2) reconstructed the Old Chinese form for shǔ 鼠 as *nhaʔ (* means a "reconstructed form") and cited affricate initial forms in the Proto-Min form $*tš^hy^B$, etc. Although he writes that the etymology is not clear, he noted Proto-Mon (a branch of Austroasiatic) *[c/s]naaʔ for "shrew," further mentioned Proto-Mon *kniiʔ "rat, mouse" and Proto-Northern-Bahnaric (a branch of Austroasiatic) *kane "rat," inferring it as related to Old Chinese and Proto-Kam-Sui (a branch of Kra-Dai) $*k^h$-no^C "rat," because occasional shifts from a front vowel to Old Chinese *a are observed.

LAAA-1 (p. 34, written by Nakazawa) compared the Japonic old form *ne* "rat, mouse" with *h-nū* "rat, mouse" in Thai as well as *hnaʔ in Old Chinese by Schuessler's reconstruction, while amending *nh* to *hn*. Thus, it was inferred that the Japonic form was borrowed from these languages, because the change *n->hn-* (">" means "changed to" and "<" means "changed from") is unnatural.

It is also possible to relate the Korean form *tʃwi* to those starting with affricates in Chinese dialects.

In Japanese and neighboring regions, a euphemism is observed; for example, *oyabito* "parent," *yome* "bride," *yomono* "night creature" in Japonic, *haozi* 耗子 "exhausting creature" in Sinitic, and *sɔsæŋwɔn* "mouse bureaucrat" in Korean. This is because mice were referred to by taboo words after rice cultivation, when the damage caused by them became severe (LAAA-1, p. 34).

Chicken

Chicken (*Gallus gallus domesticus*) originates from Southeast Asia (Lawal et al. 2020). Genetic studies suggest that the chicken expanded from East, Southeast, and South Asia to the rest of world via ancient Egypt, Syria, and Greece (Perry-Gal et al. 2015) (Fig. 11.3).

Compared with mouse/rat, it is remarkable that many word forms start with *k*-like sounds for chicken. It is onomatopoetic if they are reduplicated, such as the *kakara* type in Tungusic, the *kokor* type in Nilo-Saharan, or the *kerk* type in Iranian, although two *k* appear in a syllable, as well as *dadʒa:dʒ* and *dʒida:d* types in Semitic, in which *dʒ* or *d* sounds appear twice.

Fig. 11.3 Chicken

Schuessler (2007:292) reconstructed Old Chinese for *ji* 雞 as *kê and recognized it as an area word (=Wanderwort), listing Proto-Tai (a branch of Kra-Dai) *kəi^{B1} "chicken," Proto-Kam-Sui (a branch of Kra-Dai) *kaːi^{5}, Kadai (a branch of Kra-Dai) *ki^{A}, as well as Proto-Miao-Yao (or Hmong-Mien) *kai, and Proto-Viet-Muong *r-ka "chicken," commenting "since its origin is onomatopoetic, phonological correspondences are not regular." If chickens were domesticated in Southeast Asia, *kê, the Sinitic noun denoting "chicken," should also come from a language spoken in that region.

Numerous words for "chicken" in Asian and African languages likely originated from onomatopoeia. It is possible that Type *k-* is linked to the onomatopoeic representation of the chicken's cry, whereas Type *t-* could have been derived from another type of onomatopoeia produced by the chicken. Therefore, the lexical similarity between these word forms implies no connection between the languages and their respective groups, unless cognates can be identified.

Another type of word formation is adopting the words for "bird" or a general term for avian species, as seen in Japonic *tori* and Tibeto-Burman (Tibetic) *bya*. This is an example of a semantic transfer.

Gender differentiation for "chicken" is identified in Indo-European, Caucasian, and Turkic languages through the use of different lexical roots, which vary further in the sub-branches. Using different roots suggests that gender classification plays an indispensable role in the respective linguistic area, as shown, for example, by the widely accepted domestication of chickens for speakers' subsistence since ancient times. See "rice" for the relationship between lexical branching and rice cultivation.

Horse

The horse (*Equus ferus caballus*) was domesticated in Central Asia, especially in the lower Volga-Don region, around 3500 BCE (Librado et al. 2021). Their swift global expansion can be attributed to their application to transport and chariotry (Fig. 11.4).

A series of words including "m" are found in successive areas of eastern Eurasia. Polivanov (1927 [1974]: 155) cited that the horse was first brought to the Gobi Desert and that the Chinese borrowed the horse with its name from the Turko-Mongolian-Tungus tribes, which then spread to southern peoples. Concretely speaking, he postulates the following phonetic process: *morin* in Mongolian is related to Korean *mar-*, of which the vowel had been ʌ < ø. It is connected with Proto-Chinese *mra, also with Japanese *mma. Ramstedt (1949: 138) also recognized that the etymology of Korean *mar* came from Mongolic *morin*, concluding that "From the primary *mor* the diminutive **morqai*, **morqa* was also used and this may have been the basis for Gothiac(sic) *marka* "horse," whence German *Mähre*, English *mare* as old feminine formations."

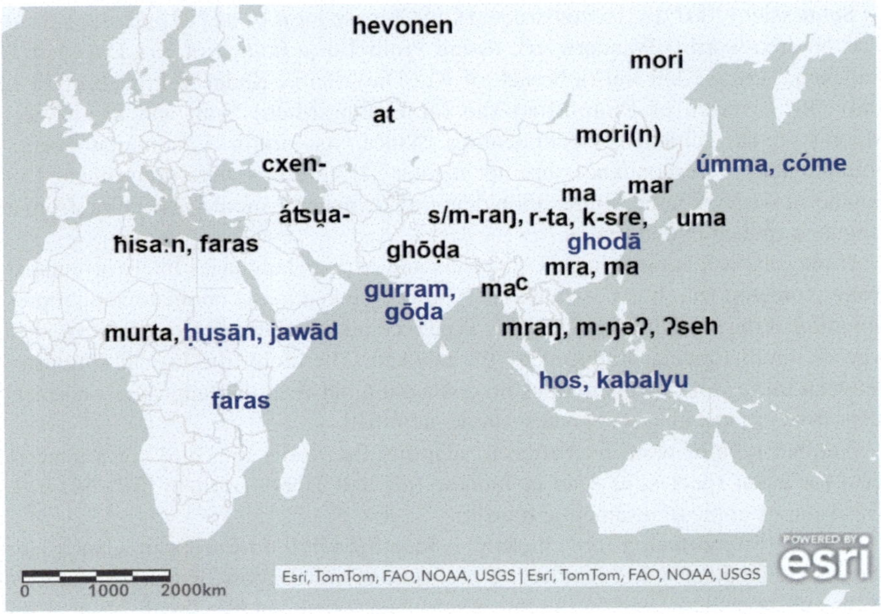

loanword

Fig. 11.4 Horse

Cincius (1965) found no trace of *-in* at the end of the forms for “horse” among Tungusic; hence, she concluded that it was added later, except for *morin* in Manchu, which was borrowed from Mongolian. Starostin et al. (2003: 946) found that the Korean word is difficult to explain as a borrowing from Mongolian.

Sagart (1999: 195–6), referring to the archeological dating of the arrival of the horse in China (ca. 2200–1800 BCE in Gansu, 1400–1100 BCE in Anyang), further postulated that “the Qijia culture (possibly to be equated with speakers of Proto-Tibeto-Burman) acted as the intermediary between the pre-Tokharians on the west and the pre-Chinese on the east,” and that “the Chinese name of the horse: ma_3 馬 *^{a}mraʔ > mæX has no apparent etymological connections within Chinese, strongly suggests that it is a loan-word.”

The horse spread to northern Korea in approximately the second century BCE, to southeastern Korea in approximately the first century BCE, and to Japan at the end of the fourth to fifth centuries AD after the Koguryo conquest (Isahaya 2019). Endo (2019) postulated that Japanese *uma* came from Chinese, not from Mongolic, Tungusic, or Korean, since the word form does not reflect the ending *-r*. Endo (2022) found that the readings of Chinese characters reflected a drastic drop in ending sounds in the oldest layer of the Korean peninsula before Sino-Korean, dating to the time when the horse spread to this area. Nozawa (2009) determined that the horses in the Japanese Archipelago were brought from Southern Korea, contradicting the possibility of a route through China. Tozaki (in this volume) revealed that the Taishu breed (in Tsushima Island at the gate of Japan from Korea) and the Noma breed (in Ehime Prefecture) first split from the Mongolian breed, which supports the archeological findings.

The Kra and Li branches of Kra-Dai show the *ŋa*-type (*ka* in Li experienced denasalization), which may be related to the Vietic *m-ŋəʔ type.

Some words are notable in the Tibeto-Burman language. First, Benedict (1972: 43) divided *m-* as a prefix meaning "high" based on Trung *mraŋ* "high, long." Sagart (1999: 196) supported this idea as follows: "However, this may be, in early Chinese sources the horse is associated with the western neighbors of the Chinese, especially the Qiang 羌 who were probably Tibeto-Burman speakers, and it would make sense if the Chinese word ma_3 馬 *ᵃmraʔ was borrowed from an early Tibeto-Burman language, perhaps one in which Benedict's *m-rang "horse" was reflected as [mrã], if we suppose that vowel nasalization was lost in Chinese (there being no nasalized vowels in the language)." Another form is derived from *r-ta (STEDT #5), found in Tibetic and its neighboring languages, and has a similar sound shape to Nilo-Saharan #murta ("#" here means "heuristically reconstructed [Nilo-Saharan] root"; LAAA-1, p. ix), although their relationship is unclear. In addition, Proto-Karen *k-sre^{T} (STEDT #7353) is similar to Austroasiatic *seh* (Shorto 2006: 531).

As for the *márkos "horse" in Indo-European, as mentioned by Ramstedt, it is found only in a few Germanic and Celtic languages. As Janhunen (1998: 425–6) pointed out with the other problems, he concluded that it is "most probably simply accidental." Leonard (2021) proposed that the Proto-Indo-European has multiple roots for "horse," with some scholars suggesting that the *m form was originally used for chariot warhorses's horse (Beckwith 2009: 397).

There are examples of loanwords from influential languages such as Arabic and Sanskrit. For example, in Africa, Arabic words are borrowed; in Nilo-Saharan, South Sudanese languages attest to loanwords from Sudanese Arabic, such as *ḥuṣān* or *jawād*; and in Bantu, a loanword from Arabic, *faras*, is used in almost the entire area of Eastern Bantu.

In Dravidian, the *gurram*-type form used in South Central and Central Dravidian languages is possibly borrowed from Sanskrit *ghōṭa*. Northern Dravidian languages borrowed the *gōḍa* type form, a modern reflex of the same form. Tibeto-Burman languages in northeastern India also have the *ghōḍa type.

The Austronesian languages have loanwords from English and Spanish/Portuguese, the *hos* and *kabalyu* types, respectively. The languages with no name for "horse" are located in the peripheral areas of the Kalahari Basin Area and Chukotko-Kamtchatkan.

Dog

The dog is considered one of the earliest animals domesticated from its ancestor, the wolf (*Canis lupus*). Domestication of the dog (*Canis lupus familiaris*) is believed to have occurred in Eurasia, although the exact location of its origin remains uncertain (Perri et al. 2021) (Fig. 11.5).

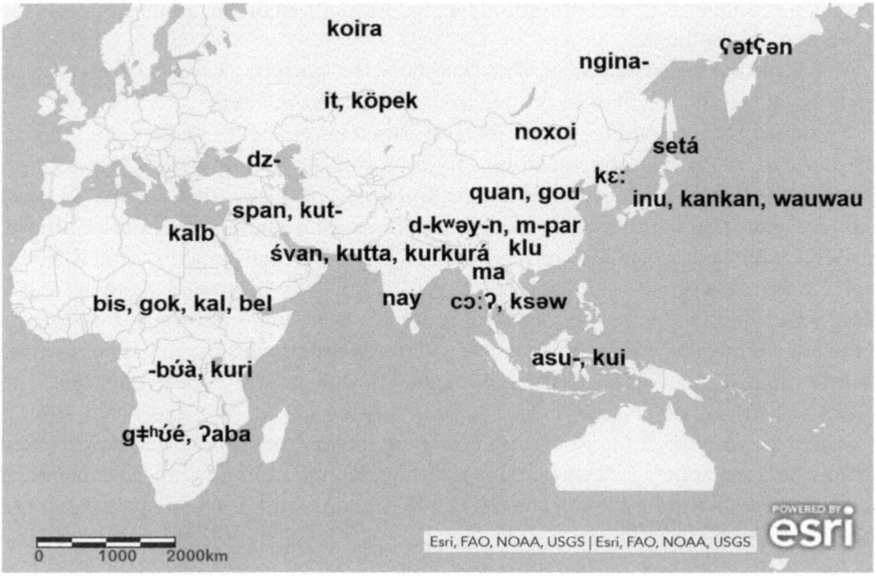

Fig. 11.5 Dog

Although word forms are different as a whole among Asian and African languages, the tendency to start from a *k*-like sound is common. It is possible to imitate the sound of barking among these languages, a type of onomatopoeia, although only one consonant is involved. Blasi et al. (2016: 10820) detected that *s* is frequently used for the word "dog" among world languages, although it is not notable in Asia and Africa.

Most Tibeto-Burman languages use forms derived from Proto-Tibeto-Burman *d-kʷəy-n (STEDT #1764). This is a cognate of the Old Chinese root *khwĭnʔ; its descendent is now a Sinitic minority. Schuessler (2007: 437) considers that *-n* in Chinese has a nominal function. Another Proto-Tibeto-Burman form is derived from *m-par ⋉ pra "WILD DOG/WOLF" (STEDT #6043) ("A ⋉ B" denotes "A and B are members of the same word family" or "A and B are co-allofams of a single etymon"; Matisoff 2003: xxxi). Languages using the root for "wolf" are spoken in southwestern China.

Schuessler (2007: 257) referred to the theory of Haudricourt, Norman, and Mei that *gǒu* 狗 is a loan or substrate word from Proto-Miao-Yao *klu^{2} "dog" which in turn is Austroasiatic: Proto-Mon *clur. He further mentioned Eberhard's hypothesis that the dog plays a significant role in Yao mythology and customs in the ancient state of Zheng in Henan (adjacent to the Shang and Zhou dynasty heartland) suggests that their inhabitants originally belonged to the Yao culture. According to Sagart (1999: 190–1), the Proto-Western-Malayo-Polynesian *u(ŋ)kuq "dog" corresponds exactly to Old Chinese 狗 *kuʔ* "dog, puppy."

Wolf

Wolves (*Canis lupus*) can be found throughout Eurasia and North America and are categorized into over thirty subspecies (Fig. 11.6).

The form for "wolf" is frequently the same as or derived from that for "dog." The euphemism or honorific word for "wolf" originated from human life, implying that humans had to live with wolves, despite being afraid of them, as shown in the word forms for "bear."

In Tibetan languages, the literary form for the word "wolf" is *spyang ki*. However, oral forms in many languages from the eastern Tibetosphere suggest that the second syllable could be linked to *khyi,* "dog," resulting in various phonetic variations of the term for "dog." If this analysis is accurate, it can be concluded that Tibeto-Burman languages predominantly employ the compound type that is widespread throughout eastern and central Eurasia.

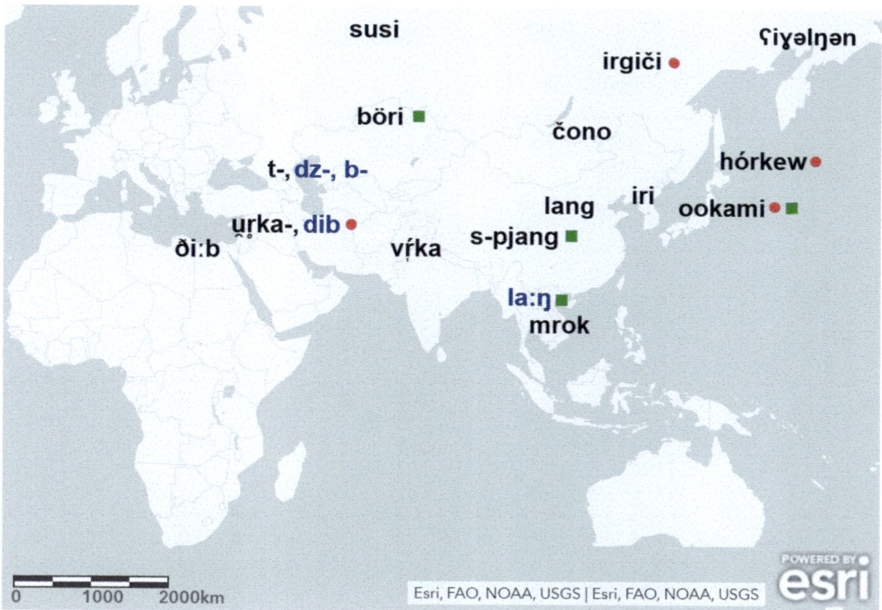

loanword • euphemism, taboo word, honorific ■ "dog" compound

Fig. 11.6 Wolf

Bear

Various types of bears (*Ursinae*) inhabit Asia and Africa, including the brown bear (*Ursus arctos*), polar bear (*Ursus maritimus*), Asiatic black bear (*Ursus thibetanus*), sun bear (*Helarctos malayanus*), and sloth bear (*Melursus ursinus*). In some languages, the terminology for "bear" differs according to the species. The main focus here is on the Asiatic black bear (Fig. 11.7).

It is noteworthy that words distinguish between black and brown bears in Tibet and Northern India, where both types of bears exist. Similarly, Chukotko-Kamchatkan distinguishes between brown and polar bears.

loanword • personification, honorific

▲ euphemism related with "dog," "monkey," or "rat"

□ distinction between "brown bear" and "Asian black bear"

■ distinction between "brown bear" and "polar bear" ◆ distinction by sex and age

Fig. 11.7 Bear

Zhang (2019) cited three forms in Old Chinese: 羆 *BAY (Proto-Tibeto-Burman) "brown bear," 熊 *WəM (Proto-Sino-Tibetan) "(Asiatic) black bear," and 能 *Nə (Austro-Asiatic). While the form 羆 did not survive in modern mainstream Sinitic dialects, *gouxiong* (the literal meaning is "dog-bear," because black bear is smaller than brown bear) served for the Asiatic black bear; thus, the two types were distinguished as "bear" and "dog-bear." However, this distinction has now disappeared in many dialects, although the Japonic *hi* (< *pi*) in *higuma* (brown bear) might have maintained this practice.

Matisoff (2003: 299–300) noticed that Japanese *kuma* and Korean *kom* are both close to the Old Chinese form, *g'ium*. The reconstruction of the Old Chinese by Baxter and Sagart (2014) is *C.[ɢ]^{w}(r)əm. The *-m* ending predates the Old Chinese form, which changed to *-ŋ* in Old Chinese. The initial *g'* or *ɢ* changed to *ɦ* in Middle Chinese around the seventh century AD, further changing to *h*, *ɕ*, *j*, etc. in Modern Sinitic dialects. Frellesvig and Whitman (2008: 36) cite the correspondence of *kom* in Middle Korean and *kuma* in Old Japanese as examples of raising the mid vowel *o > *u*. As for the deletion of the final vowel (apocope), as seen in Middle Korean, there are parallel examples. Thus, the proto form can be reconstructed as *koma.

According to Schuessler (2007: 542), the Old Chinese final *-m of *xiong* is preserved in Min dialects: Amoy *him*^{A2}. He also noticed that Sino-Tibetan forms, such as Proto-Tibeto-Burman *d-wam changed to Written Tibetan *dom* "bear," Proto-Loloish *k-d-wam, and written Burmese (*wak-*) *wam*, and other languages in East Asia have similar words. Middle Korean *kwom*, Old Japanese *kuma*, Ainu *kamuy* "bear;" Taiwan Austronesian languages include *cumai/cumay*; and Proto-Monic (Austroasiatic) *kmum is "Himalayan black bear." Schuessler (2007: 398) writes that *neng* (Old Chinese *nə̂) "a kind of bear" apparently is an area word of Austroasiatic origin, for example, Kharia *bɔnɔi* "bear."

In Tibetic, the common form *dom* for "black bear" is derived from the Proto-Tibeto-Burman *d-wam "BEAR" (STEDT #2777). Some Tibetic varieties possess another root *dred* for "brown bear." The double-prefix form *k-d-wam is reconstructed in Proto-Loloish, which is relevant for words with a *k*-initial consonant used in various East Asian languages.

As for the Wanderwort found among Eastern Asian languages, the word family within a particular language is an important cue for determining the donor language. Zhang (2019:10) recognized that some words meaning "dark" comprise a word family with *xiong* 熊 "black bear." In contrast, Janhunen (2003: 2) considers that Japanese *ku.ma* came from *ku.ro* "black;" hence, Korean *gom* is a Japonic borrowing in Pre-Proto-Koreanic. Sakakura (1982) sought the etymon of *kuma*, which originated from the verb *kumu* "to hide," and gave the general meaning as "hidden matters or places," for example, like a valley in the heart of a mountain, corners in shadow, and gods, which have alternate the sound of *u* and *a* with the ordinary word for gods *kami*. The Ainu word *kamuy* for "bear" also means spirit deities due to synecdoche or a type of belief in a bear (LAAA-1, p. 242). Thus, there are at least two candidates for the donor language: Sinitic and Japonic. According to recent theories, the Japonic language once existed in northeastern China.

Loanwords are used in Kra-Dai (*jong* from Sinitic), Mongolic (*xar görȭs* "black wild beast" from Chinese 黑熊 "lit. black bear" and *noxo xar görȭs* "dog black wild beast" from Chinese 狗熊 "lit. dog(-like) bear"), and Iranian (*melu* possibly from Hindi). These are from neighboring languages, but Austronesian languages use *ber*, borrowed from English.

Tungusic and Turkic languages have words for "bear," denoting "grandfather or old man," which is a personification. Ainu uses *kamuy* for bear, which is an honorific. Terms for "dog," "monkey," or "rat" are used with "bear" in Sinitic, South Asia, and Dravidian, respectively, which might be euphemisms.

Rice Plant

Edible rice (*Oryza sativa*) has two main subspecies: *japonica* and *indica*. The *japonica* variety is typically grown in temperate East Asia, the uplands of Southeast Asia, and the highlands of South Asia, while the *indica* variety is generally grown in the lowlands of tropical Asia (Chang 1976; Wang et al. 2018) (Fig. 11.8).

Despite continuing debates, China has been determined to be the origin of domesticated rice species in Asia. Early rice cultivation has been documented across

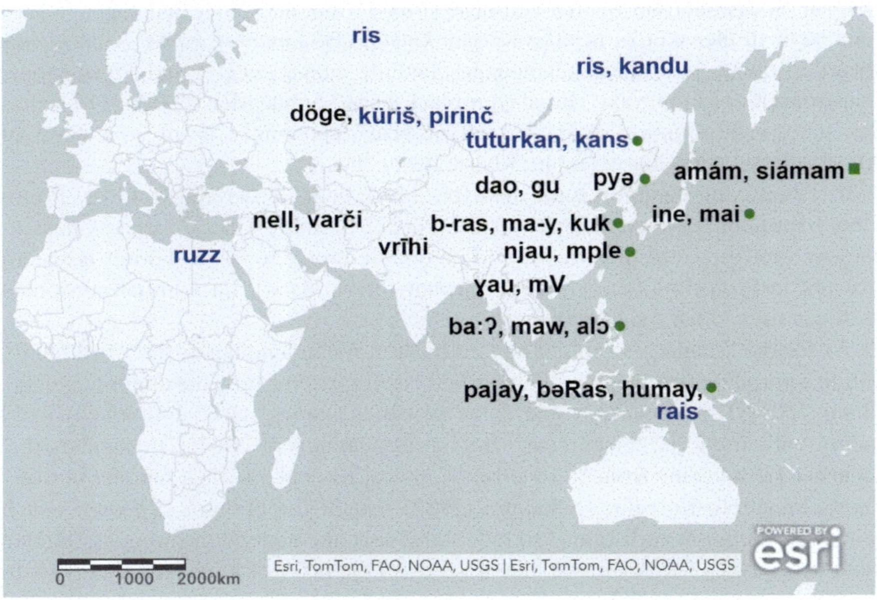

loanword • rice plant/rice grain distinction ■ plant including millets

Fig. 11.8 Rice plant

many cultures, including China, India, and Southeast Asian civilizations. Nonetheless, the oldest archeological evidence is found in central and eastern China (Huang et al. 2012; Wang et al. 2018), dating back to 7000–5000 BCE. Miyamoto (2017: 4) cited the dating as 12,060 cal. BP is in the Yuchanyan site in Hunan, China, and 12,830 cal. BP in Poyang, Jiangxi, China. Rice cultivation was found in the Shandong peninsula as late as circa 2500 BCE, and the Liaodong peninsula in about 2500–2000 BCE. After 1500 BCE, irrigated wet-rice agriculture appeared in Southern Korea and then in northern Kyushu, Japan, around the ninth to eighth centuries BCE. The agricultural complex entered northern Kyushu, Japan.

During the second millennium BCE, rice cultivation was carried out as part of agricultural activity in the Kashmir and Harappan regions India (Smith and Frederiksen 2000). Rice cultivation in Southeast Asia can be traced back to Indonesia, which may have started over 1500 years ago. The Philippines also cultivated rice from the first to second millennium BCE, while Thailand dates back to 2200 BCE.

Schuessler (2007: 207–8) reconstructed the Old Chinese *lûʔ for *dào* 稻 in Sinitic and concluded that its etymology is uncertain. He cited similar-looking words, such as Proto-Miao-Yao *nblau^{A} sic and Proto-Viet-Muong (Austroasiatic) *ʔa-lɔːʔ "unhusked rice," Vietnamese *lúa* "paddy," and Khmuic *lɔʔ "glutinous rice," whose meaning seems to be connected to the Mon-Khmer (Austroasiatic) etymon with *nuo* 糯.

Sagart (2011) argues that the character *dao* 稻 originally denoted "rice grains" before shifting its meaning to "rice plant" based on evidence in Zhou bronze inscriptions. He also provides a potential Old Chinese reconstruction of *dao* as *m.lˤuʔ, which may support a connection with the proto-Hmong-Mien word *mbləu^{A} "rice plant." This suggests that the Hmong-Mien borrowed the "rice plant" form after the Chinese had changed the meaning of *dao* 稻 (Sagart 2011: 128).

The Tibetic form *'bras* "rice, rice plant" is widespread throughout the Tibetosphere, and its origin has been long debated. Sagart (2011: 127) argues that it is etymologically related to Proto-Austronesian *bəRas, which, as well as *'bras*, may have meant no more than "fruit; dehusked grain of cereal." Tibetic languages spoken in rice-cultivating areas possess word forms other than *'bras*, e.g., *drus ma* "polished grain of rice" (Suzuki 2022: 265–273), which originally denoted "millet" (Tournadre and Suzuki 2023: 839). In these languages, *'bras* no longer means grain, but is a general term for "rice" or plant, suggesting they created more precise semantic categories for rice after *'bras* was used as a general term for "rice."

Ferlus (2010) extends the discussion on the proto-forms for "rice" in Proto-Austronesian *bəRas, Proto-Austroasiatic *C.rac, Old Chinese *m.rats (糲 *li*; not 稻 *m.lˤuʔ as provided by Sagart 2011), Tibetan *'bras*, Sanskrit *vrīhi*, and Latin *oryza*. The sound components are scrutinized in a quadripartite manner. Ultimately, the analysis concludes that "the pre-Neolithic populations of India consumed the wild *indica* obtained by gathering, as well as the rice cultivated; its name might have come from Austroasiatic populations from China before the Munda expansion to India" (Ferlus 2010: 70).

Matsumoto (2011 [2016]: 241–243) interpreted the etymology of the Japonic *ine/ina* and *yone/yona* "rice plant" as *yə "good" plus *na "food," meaning "good food." As for *ko2me2* "rice grain," he observed that combining the vowels "*o2* + *e2*" is rare in Old Japanese, revealing its heterogeneity; hence, its emergence is late, probably because it is a loanword from Austronesian *humay*. That means there was no distinction of words between rice plant and rice grain in Japanese. Also, *pyɔ* "rice plant" in Korean was related to *piye* "barnyard millet" in Old Japanese, and *piyapa* "barnyard millet" in Ainu, further raising the possibility that the word and plant were transmitted to the Korean peninsula meaning crop in general. Nonetheless, in our opinion, *pie* and *pyə* are possibly related with *bai* 稗 "barnyard millet" in Sinitic.

Whitman (2011: 155–156) reconfirmed that no common rice vocabulary exists among Japonic, Koreanic, and other northern languages. He further interpreted this as evidence that the Japonic and Koreanic migrated to northeast Asia and the Korean peninsula before the advent of wet-rice agriculture. While Matsumoto (2011 [2016]: 243–244) gave another interpretation of this absence from the viewpoint that only rice cultivation technology was adopted by the Japanese independently without immigrants from the Continent. Nonetheless, this explanation contradicts recent ancient DNA studies that have identified totally new DNA that entered the Japanese Archipelago after the Yayoi period.

The word forms related to English *rice* in most European languages come from the south, originating from Greek *ὄρυζα* (*oryza*). This form expanded into Arabic *ruzz* (Classical Arabic *ʔuruz*, *ʔurz*, and *aruz*) (LAA, p. 92), which was, unusually, borrowed into languages spoken in the Iberian Peninsula, such as Spanish *arroz*. The Greek form was borrowed in some Kartvelian languages and Armenian; however, it probably came from ancient Iranian languages (*birinj*), which have the same etymology as Sanskrit *vrīhi* (LAA, p. 89).

Wheat

Wheat, along with corn and rice, was among the first crops cultivated by humans approximately 10,000 years ago. Wheat cultivation originated in central Eurasia, stretching from the Caucasus to Mesopotamia (Zohary et al. 2012) and subsequently developed across the Mediterranean, spreading to Europe and Asia (*Nipponica*) (Fig. 11.9).

Starostin et al. (2003: 935) reconstructed their Proto-Altaic form as *miurgu for wheat, based on Proto-Tungusic *murgi, Japonic *mùnkí, and Korean *mírh. They cited Manchu *muǯi*, Jurchen *mir-ɣe-i* "product of agriculture," and so on. At the end of their description, they cited Turkic **bogu-* (<**borgu-*?) in **bogu-daj* > **bugdaj*, which in turn borrowed in Mongolic *buɣudaj*.

Schuessler (2007: 374) links his Old Chinese form *mrêk for *mai* 麥 "wheat" with Sino-Tibetan *mrə(k) > Written Tibetan *bra-bo* "buckwheat," in addition to Proto-Lolo-Burmese *g-ra^2 "buckwheat." According to Zhengzhang (1990: 71), the

loanword ◆ "rice + wheat" compound

Fig. 11.9 Wheat

voiced ending of Japonic *mugi* is treated as a trace of Old Chinese voiced endings of the entering tone.

Watanabe (2010) infers that the Chinese character in the Yin Dynasty did not distinguish wheat from barley. Furthermore, if wheat was brought to China after the Zhou Dynasty, then the Chinese character would have denoted barley.

It is worthwhile to consider some possibilities for future scrutiny. First, the Proto-Tibeto-Burman *m-grwa "WHEAT/BUCKWHEAT" (STEDT #5858) may be connected with the Old Chinese *m-rˤək, with a metathesis of *g* and the zero final: **m-rwa-g (** means a "hypothetical form developed from a reconstructed form"). Second, the Middle Korean form *mirh H* (*H* stands for high tone) can be interpreted as a loanword from the Middle Chinese *mie lok* 弥勒 of "maitreya," since people worship the value of this crop. The ending *-h* in Middle Korean is derived from *-k* in Old Korean. It is also comparable to "rice plant" in Korean being rendered as *posar* 菩薩 "maitreya" in the *Jilin leishi* 鶏林類事 in the twelfth century, which changed to *psʌr* in Middle Korean around the fifteenth century, and *ʔsar* in Modern Korean. However, Ogura (1943 [1975]) interpreted its etymology as *pyɔsir* "bureaucracy." Also, the tone in Middle Korean is high; if it were contracted from two syllables in Old Korean, the tone would be rising in Middle Korean. Another possibility is a metathesis of r in the Old Chinese *m-rˤək (proposed by Masaki Nohara, p.c.).

There are loanwords from Chinese around the Sinitic-speaking area, Japanese in Ainu, Russian in northern Eurasia, and Indo-Aryan or Persian in southern Eurasia. Additionally, there are two types of compounds in the map of "wheat"; one is the "X + wheat" compound in the northern Sinitic-speaking area and Japan (in Japanese "wheat" and "barley" share a form: ex. *ko-mugi* in which *ko* means "small" and *o:-mugi* in which *o:* means "large") and the "rice-wheat" compound in Kra-Dai and Ryukyuan ("wheat," "barley," and "rice" share a form).

Broomcorn Millet

According to Lu et al. (2009), broomcorn millet, also known as proso millet or common millet (*Panicum miliaceum*), was domesticated around 10,000 BP in Northern China. Millet farming in China expanded from the Yellow River area to Southwest China starting with the Neolithic period (d'Alpoim Guedes et al. 2013; Tao et al. 2023) (Fig. 11.10).

Schuessler (2007: 472) reconstructed his Old Chinese form for *shǔ* 黍 as *nhaʔ, adding "?" and "perhaps," and cited the Old Chinese reconstructed form by Baxter

loanword **forms shared by other millets**

■ grain including "rice plant" and other millets

Fig. 11.10 Broomcorn millet

*hjaʔ. He compared this root with the Written Tibetan *nas* "barley," adding "perhaps," probably meaning his reconstruction was based on this comparison in reverse.

Three main lexical items have been confirmed for the millet series in Tibeto-Burman: Proto-Lolo-Burmese *tsap ("MILLET"; STEDT #5861), Proto-Lolo-Burmese *C-lu-k ("MILLET"; STEDT #5860), and Proto-Tibeto-Burman *khrəy ("MILLET/RICE"; STEDT #5867). *tsap and *C-lu-k are chiefly associated with the *Panicum* and *Setaria* species, whereas *C-lu-k and *khrəy are mainly linked to the *Setaria* and *Echinochloa* species. The lexical diversity of Tibeto-Burman millet species implies uncertainty in distinguishing between these species in language usage. For instance, Tibetic uses the single word *khre* for "millet," without specifying the species within the millets, like English *millet*. However, the Tibeto-Burman etyma do not correspond to Sinitic in their current usage.

The newly coined Japonic term *kibi* or *kimi* originally denoted "high millet, great millet, corn, and sugar cane" (LAAA-2, p. 43). Cognates are not found in other languages.

The maps of millets show a tendency for some words to be shared by other millet species in the area surrounding South Asia. On the other hand, the three millets are distinguished by different names in East Asia, providing evidence that millet was domesticated in East Asia and spread to surrounding areas.

Ainu have the word *amam* for "grains," including rice plants and millets. They use a compound word to specify the grains of each species.

Foxtail Millet

Foxtail millet (*Setaria italica*) was domesticated approximately 16,000 years ago. It became a recognized crop 9000–10,000 years ago and became popular in Northern China approximately 5000–6000 years ago (Diao and Jia 2017) (Fig. 11.11).

The most widespread root for "foxtail millet" in Sinitic, *su* (reconstructed as *sjək), denotes "foxtail millet grain." Sagart (2011: 124) provides the etymon for "foxtail millet plant" as *ji* 稷 (reconstructed as *[ts]ək). The term for foxtail millet shares some cognates with Tibeto-Burman, such as Written Tibetan *chag* "dry fodder for horses and other animals." Sagart (2011: 124) argues that rice and foxtail millet were cocultivated between the Yangzi Valley and the Yellow River valley. Rice has been cultivated since approximately 7000 BCE in the former, while foxtail millet was domesticated in approximately 6500 BCE in the latter as one of the early domesticated plants in Northern China. Therefore, Sinitic languages spoken north of the Yangzi Valley already distinguished between the terms for grain and the plant of foxtail millet on a lexical root level. Matsumoto (2011 [2016]: 236) found a word family derived from *dzjək 食 "to eat": *tsjək 稷 "foxtail millet," *sjək 粟 "foxtail millet," and *sjo 黍 "broomcorn millet," observing that this is consistent with the staple food in the Yellow River basin being millet, not rice.

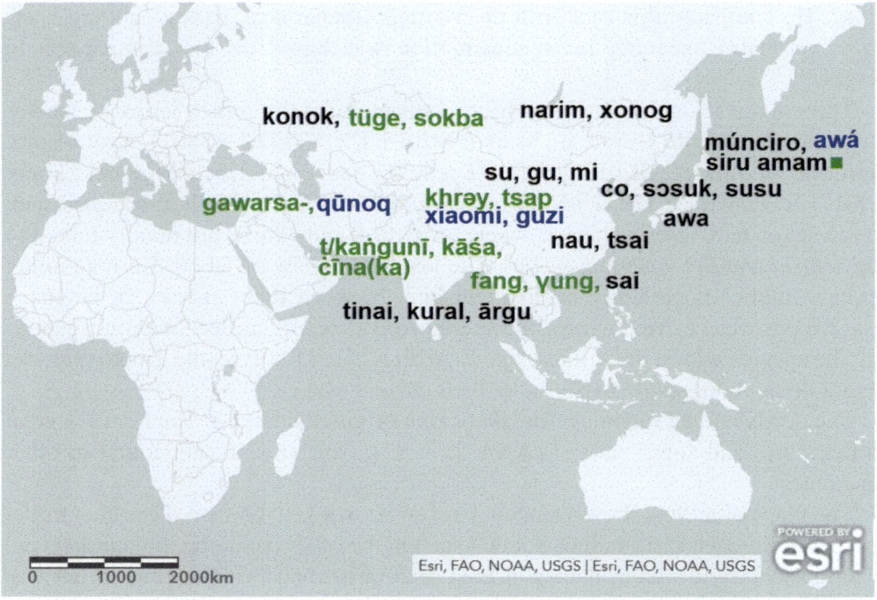

loanword **forms shared by other millets**

■ grain including "rice plant" and other millets

Fig. 11.11 Foxtail millet

The Japonic term *awa* (<Old Japonic *apa*) is used for both the plant and grain of foxtail millet. This term may be related to Proto-Turkic *arpa "barley", Mongolian *арвай* "barley" (Mongolic) and Manchu *arfa* "barley, oats" (Tungusic) (LAAA-2, p. 68).

Barnyard Millet

The term "barnyard millet" in English refers to two *Echinochloa* species, namely *Echinochloa esculenta* or "Japanese barnyard millet" and *Echinochloa ramentacea* or "Indian barnyard millet." While the former is believed to have originated from East Asia and is extensively cultivated, the latter is grown in South Asia; however, when it was domesticated remains uncertain (Fig. 11.12).

Schuessler (2007: 155) observed that the etymon of *bai* 粺 sic Old Chinese *brêh is close to *li* 糲 *ras "coarse grain." He compared it with Austroasiatic *pih "rice" and Tibeto-Burman: Mru *rai* "a rice species," Lushai *ṭai*^{F} < *traih* (< *-s) "species of early rice."

In Tibeto-Burman languages, based on limited data, a few languages use words derived from Proto-Lolo-Burmese *C-lu-k ("MILLET;" STEDT #5860) and Proto-Tibeto-Burman *khrəy ("MILLET/RICE;" STEDT #5867).

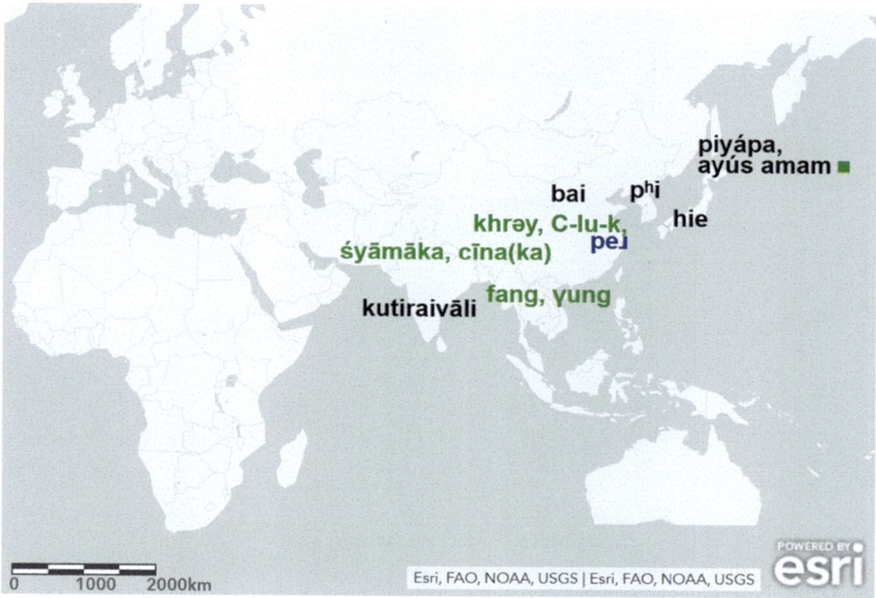

loanword **forms shared by other millets**

■ grain including "rice plant" and other millets

Fig. 11.12 Barnyard millet

Taro

Taro (*Colocasia esculenta*) is believed to have originated in Southern India and Southeast Asia and is now grown in tropical regions (Fig. 11.13).

Schuessler (2007: 589) treats *yu* 芋 Old Chinese *wah as an area word (Wanderwort), listing Proto-Miao-Yao *vəu^{B2} and Written Burmese *wa*^{C} as a "kind of potato."

The Tibeto-Burman languages have several roots for "taro," which also denotes "yam" as proto-lexical forms. In the STEDT database, PTB *m-n(w)ay "YAM/TARO" (STEDT #5598) and *kywəy "YAM" (STEDT #2381) are the principal forms.

Blench (2012: 24) identifies five common Indo-Pacific roots of "taro": #trawʔ (Austroasiatic), #tales (Austronesian), #ma (Papuan, Austronesian), #biRaq (Austronesian, Tai-Kadai), and #poŋ (Austroasiatic, Sino-Tibetan) (# here means "quasi-reconstruction"; Blench 2012: 21). His analysis (2012: 36–37) shows that the word forms #trawʔ (Austroasiatic) and #biRaq (Austronesian, Tai-Kadai) were not only centers for domestication but also engines of language phylum expansion from the Middle Mekong area and Taiwan, respectively.

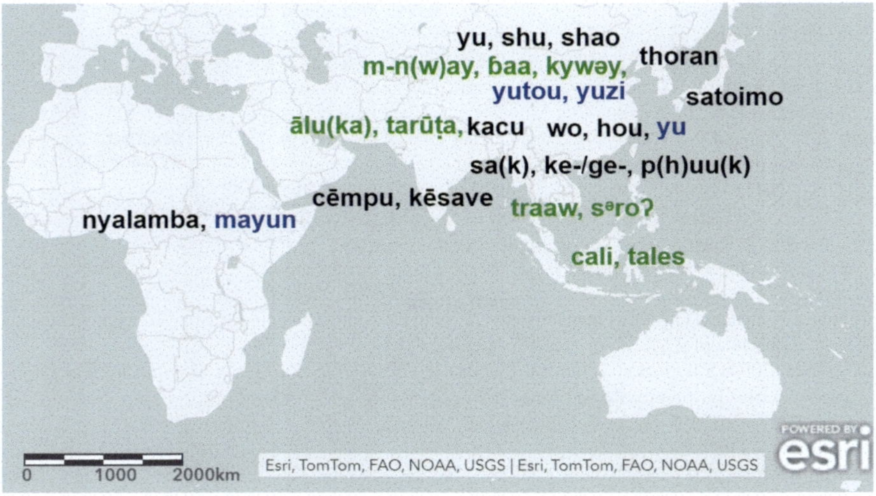

loanword **forms shared by yam**

Fig. 11.13 Taro

The fundamental Japonic word for “taro” and “yam” is commonly *imo*, which mainly denotes “potato” in the modern period; however, etymologically, it seems that *imo* denoted “taro.”

Yam

Yam refers to several species of the *Dioscorea* plant. The main word for “yam” in Sinitic is *shu*薯, which is lexically distinguished from *yu* 芋 “taro.” However, it is difficult to determine which species of *Dioscorea* are represented in each language data. For instance, the Japanese words *yamaimo* and *zinenzyo* were originally used to refer to different plants. However, in everyday language, not all speakers distinguish between them. In many languages, the same lexical roots are used for “taro” and “yam,” such as Japonic *imo*, Proto-Tibeto-Burman *m-n(w)ay, *kywəy, Austroasiatic *traaw*, Indo-Aryan *ālu*, and even Sinitic *shu* (Fig. 11.14).

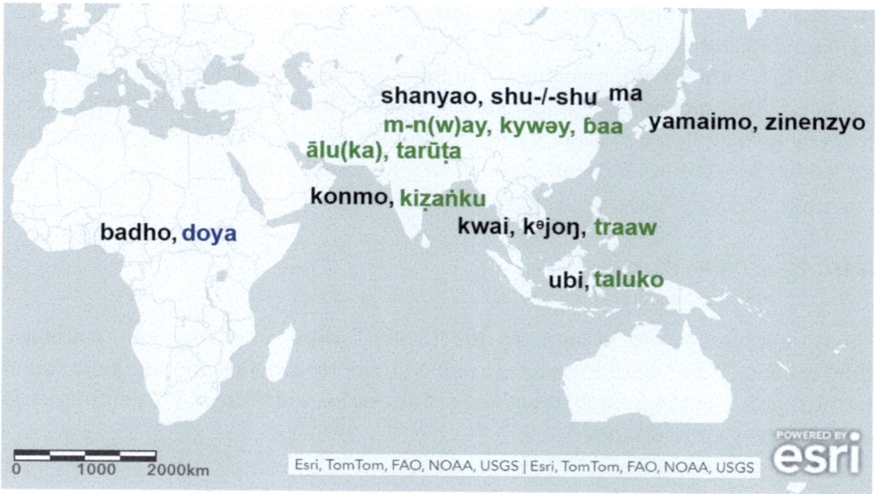

loanword **forms shared by taro**

Fig. 11.14 Yam

11.4 Discussions

All of the words treated in this chapter are new species that emerged after wild animals and crops were domesticated. At least four main types of adaptation methods were observed.

Applying the Existing Generic Term to a New Species

The typical case of this is Japanese *tori* "bird" to denote "chicken." If it is necessary to distinguish it, a determiner is added. In this case, *niwa* "garden" is added; as a result, *niwa-tori* means "chicken." This method is widely used for other cases, such as Sinitic *mai*, which originally denoted "barley" when wheat was introduced and applied to "wheat." In modern usage, if it is necessary to distinguish it, *da* "big" or *xiao* "small" are added to denote "barley" or "wheat" respectively. In Siamese (Kra-Dai), *khau*, originally denoting "rice," existed. When "wheat" started to be imported recently, a determiner *salii* was added, and *khau salii* was used to denote "wheat."

If a new species appears frequently, then a new term is used. For example, if Matsumoto's (2011) interpretation is correct, the Japonic *yo-na-i* or *i-na-i* "rice" originally did not distinguish "rice plant" and "rice grain;" only after the wide adoption of cultivation did it become necessary to distinguish between the plant and grains, and the Austonesian *humay* was borrowed to denote "rice grain." Because wheat is not cultivated as widely as rice, such differentiation of terms did not occur for wheat. As for "millet," the differentiation for broomcorn millet, foxtail millet, and barnyard millet is made with totally different words in Sinitic, Korean, and Japonic. In Southeast Asian language families, they are denoted with the same word in general. This means that millets are not important in this area. On the contrary, there is abundant horse vocabulary in pastoralism people like Uyghur, distinguishing gender, age, color, castration, and so on with totally different words.

Coining a New Word

This is clearly observed in the case of "chicken." Many onomatopoeic words use *k*-like sounds. These word forms were coined by independently imitating the sound of crowing or clucking. Such a tendency is also seen with "dog." This similarity is due to language universality.

Borrowing the Word Form from the Donor of the New Species

If a word is borrowed relatively recently, its nature as a loanword is clear. However, if its history is long, the word is felt as if it were an indigenous word, and it is called "Wanderwort." In Japonic, *me* and *ba* "horse" follow the sound laws of *Go-on* and *Kan-on*, respectively; this frequently occurs in Sinitic-origin compounds. On the other hand, *uma* is felt as if it were a Japanese word of origin for ordinary people.

Emergence of New Words with Taboo, Honorific, Etc.

Wolf and bear are dangerous animals, so people avoid calling them directly and use a different word with euphemisms. Another type is using honorific words. This type is seen in restricted areas, the northern areas for "bear" and only the Japonic and Ainu for "wolf."

Few words spread across language families in the "rat/mouse" case. As a matter of fact, this is an ordinary case because of the definition of language family. Only the third type shown above can serve as a cue for detecting the directionality of language dispersion.

11.5 Conclusion

Regarding the entry process of the agricultural complex into Japan, Miyamoto (2017: 10–16; 2019) explained his hypothesis concerning the existence of four stages in agriculture's development in northeast Asia. The first stage is circa the sixth millennium BCE, when foxtail and broomcorn millet originally domesticated in the middle and lower basin of the Yellow River spread northward to the Liaoning area, further spread to north Korea around 4000 BCE, and then to Southern Korea around 3300 BCE. The second stage involves the spread of cultivated rice. Approximately in the fifth millennium, the middle and lower basins of the Yellow River accepted rice agriculture. It reached the Shandong peninsula at the latest in circa 2500 BCE. Rice dispersed to the Liaodong peninsula about 2500–2000 BCE. The third stage is after 1500 BCE, when the Mumun culture spread southward from Liaodong to Korea. Irrigated wet-rice agriculture appeared in Southern Korea. The fourth stage is around the ninth to eighth centuries BCE when the agricultural complex entered northern Kyushu, Japan. Shoda (2009) suggested that rice, wheat, foxtail millet, and broomcorn millet entered Japan at once, but he did not provide a conclusion. He also considered that barnyard millet was already cultivated in the Jomon period. As for taro, archeologists still have not detected its trace in the Jomon period. Miyamoto (2017, 2022) further developed his four-stage theory of the migration process of the Japonic-speaking people.

It is difficult to determine the exact date when the words for these crops arrived in Japan based on linguistic evidence. However, it is surprising that the geographical distribution of these crops is very limited except for "wheat." Rice and these millets were domesticated in China. In the surrounding areas, only rough a differentiation of lexical roots is made for the minute sorts of millets or rice. These crops and words were difficult to find in the outer areas, so it was impossible to map. Paradoxically, the absence of these crops and words in the outer areas clearly carved in relief the homeland of Japonic-speaking people, as Miyamoto's theory suggests. Another surprise is that the honorific expressions for "bear" and "wolf" are found only in Japan and the neighboring regions. A similar situation was identified for "sun" (LAA: p. 19). Worship also serves as a hint to identify the roots of human groups.

Competing Interests The authors have no relevant financial or non-financial interests to disclose.

References

Baxter WH, Sagart L (2014) Old Chinese reconstruction, version 1.1. https://ocbaxtersagart.lsait.lsa.umich.edu/BaxterSagartOCbyMandarinMC2014-09-20.pdf

Beckwith CI (2009) Empires of the silk road: a history of Central Eurasia from the Bronze Age to the present. Princeton University Press, Princeton, NJ

Benedict PK (1972) Sino-Tibetan: a conspectus. Cambridge University Press, Cambridge

Blasi D et al (2016) Sound–meaning association biases evidenced across thousands of languages. PNAS 113(39):10818–10823. https://doi.org/10.1073/pnas.1605782113

Blench RM (2012) Vernacular names for taro in the Indo-Pacific region: implications for centres of diversification and spread. In: Spriggs M, Addison D, Matthews PJ (eds) Irrigated taro (Colocasia esculenta) in the Indo-Pacific. National Museum of Ethnology, Suita, pp 21–43. https://doi.org/10.15021/00002508

Chang T-T (1976) The origin, evolution, cultivation, dissemination, and diversification of Asian and African rices. Euphytica 25:425–441. https://doi.org/10.1007/BF00041576

Cincius VI (1965) Manchu-Tunguz data on the word "horse". Central Asiatic J 10(3/4):151–156. https://www.jstor.org/stable/41926729

d'Alpoim Guedes J, Jiang M, He K, Wu X, Jiang Z (2013) Site of Baodun yields earliest evidence for the spread of rice and foxtail millet agriculture to south-west China. Antiquity 87(337):758–771. https://doi.org/10.1017/S0003598X00049449

Diao X, Jia G (2017) Origin and domestication of foxtail millet. In: Doust A, Diao X (eds) Genetics and genomics of Setaria. Plant genetics and genomics: crops and models, vol 19. Springer, Cham, pp 61–72. https://doi.org/10.1007/978-3-319-45105-3_4

Endo M (2019) Geographical distribution of two loan words, "pumpkin" and "horse," in Chinese and the surrounding languages. In: Suzuki H et al (eds) Collected papers on eastern Asian geolinguistics, pp 90–98. [in Chinese] https://publication.aa-ken.jp/sag_mono6_eastern_asian_2019.pdf

Endo M (2022) The weakening and deletion of final sounds in the earliest Sino-Japanese and Sino-Korean. Keizai Kenkyu 73(4):157–164. [in Japanese]. https://doi.org/10.34321/22392

Ferlus M (2010) The Austroasiatic vocabulary for rice: its origin and expansion. J Southeast Asian Linguist Soc 3(2):61–76

Frellesvig B, Whitman J (2008) Evidence for seven vowels in Proto-Japanese. In: Frellesvig B, Whitman J (eds) Proto-Japanese: issues and prospects. John Benjamins, Amsterdam, pp 15–41. https://doi.org/10.1075/cilt.294.03fre

Huang X, Kurata N, Wei X et al (2012) A map of rice genome variation reveals the origin of cultivated rice. Nature 490(7421):497–501. https://doi.org/10.1038/nature11532

Isahaya N (2019) Higashi Ajia ni okeru uma bunka no toho tenkai [Eastward expansion of the horse culture in East Asia]. In: Migishima K et al (eds) Uma no kokogaku [Archaeology of horse]. Yuzankaku, Tokyo, pp 48–57. [in Japanese]

Janhunen J (1998) The horse in East Asia: reviewing the linguistic evidence. In: Mair VH (ed) The Bronze Age and Early Iron Age peoples of eastern central Asia. Vol. 1. Archaeology, migration and nomadism, linguistics. Institute for the Study of Man, Washington, pp 415–430

Janhunen J (2003) Tracing the bear myth in Northeast Asia. Acta Slavica Iaponica 20:1–24. http://hdl.handle.net/2115/39417

Lawal RA, Martin SH, Vanmechelen K et al (2020) The wild species genome ancestry of domestic chickens. BMC Biol 18:13. https://doi.org/10.1186/s12915-020-0738-1

Leonard SP (2021) Hipponyms in Indo-European: using register to disentangle the etyma. J Lang Relationsh 19(1–2):1–14. https://doi.org/10.1515/jlr-2021-191-206

Librado P, Khan N, Fages A et al (2021) The origins and spread of domestic horses from the Western Eurasian steppes. Nature 598(7882):634–640. https://doi.org/10.1038/s41586-021-04018-9

Lu H, Zhang J, Liu K, Wu N et al (2009) Earliest domestication of common millet (*Panicum miliaceum*) in East Asia extended to 10, 000 years ago. PNAS 106(18):7367–7372. https://doi.org/10.1073/pnas.0900158106

Matisoff JA (2003) Handbook of Proto-Tibeto-Burman. University of California Press, Berkeley. https://escholarship.org/uc/item/19d79619

Matsumoto K (2011) [2016] Ine kome no hikaku gengogaku [Comparative studies on rice]. In: Kotoba wo meguru shomondai [Problems concerning language]. Sanseido, Tokyo, pp 215–245. [in Japanese]

Miyamoto K (2017) Tohoku Ajia no shoki noko to Yayoi no kigen [Early agriculture in Northeast Asia and the origin of Yayoi]. Doseisha, Tokyo. [in Japanese]

Miyamoto K (2019) The spread of rice agriculture during the Yayoi period: from the Shandong Peninsula to Japanese Archipelago via Korean Peninsula. Jpn J Archeol 6(2):109–124. https://hdl.handle.net/2324/2236690

Miyamoto K (2022) The emergence of 'Transeurasian' language families in Northeast Asia as viewed from archaeological evidence. Evol Hum Sci 4(e3):1–17. https://doi.org/10.1017/ehs.2021.49

Nozawa K (2009) Uma [Horse]. In: Ajia no zairai kachiku [Indigenous livestocks in Asia]. Nagoya Daigaku Shuppankai, Nagoya, pp 187–214. [in Japanese]

Ogura S (1943) [1975] Ine to bosatsu [Rice and maitreya]. In: Ogura Shimpei Hakase Chosakushu [Collected works of Dr. Shimpei Ogura], vol 2. Kyoto University, Kyoto, pp 415–445. [in Japanese]

Perri AR, Feuerborn TR, Frantz LAF, Larson G, Malhi RS, Meltzer DJ, Witt KE (2021) Dog domestication and the dual dispersal of people and dogs into the Americas. PNAS 118(6):e2010083118. https://doi.org/10.1073/pnas.2010083118

Perry-Gal L, Erlich A, Gilboa A, Bar-Oz G (2015) Earliest economic exploitation of chicken outside East Asia: evidence from the Hellenistic Southern Levant. PNAS 112(32):9849–9854. https://doi.org/10.1073/pnas.1504236112

Polivanov ED (1927) [1974] Toward the question of the kinship relations of Korean and the "Altaic" languages. In: Polivanov ED (ed) Selected works. Mouton, Den Haag, pp 149–158. https://doi.org/10.1515/9783110815627-008

Ramstedt GJ (1949) Studies in Korean etymology. Suomalais-Ugrilainen Seura, Helsinki

Sagart L (1999) The roots of Old Chinese. John Benjamins, Amsterdam. https://doi.org/10.1075/cilt.184

Sagart L (2011) How many independent rice vocabularies in Asia? Rice 4:121–133. https://doi.org/10.1007/s12284-011-9077-8

Sakakura A (1982) Gogen—"kami" no gogen wo chuushin ni [Etymology, with special reference to the etymology of "kami"(God)]. In: Goi genron [Principles of vocabulary]. Meiji Shoin, Tokyo, pp 89–107. [in Japanese]

Schuessler A (2007) ABC etymological dictionary of Old Chinese. University of Hawai'i Press, Honolulu

Shoda S (2009) Tohoku Ajia no senshi noko to Yayoi noko—Chosen hanto wo chushin toshite—[Prehistoric agriculture in Northeast Asia and agriculture in the Yayoi period, with special reference to the Korean Peninsula]. In: Shitara H et al (eds) Shokuryo no kakutoku to seisan [Acquisition and production of food]. Doseisha, Tokyo, pp 39–54. [in Japanese]

Shorto HL (2006) A Mon-Khmer comparative dictionary. Research School of Pacific and Asian Studies, Australian National University, Canberra. https://doi.org/10.15144/PL-579

Smith CW, Frederiksen RA (2000) Sorghum: origin, history, technology, and production. Wiley, New York

Starostin S, Dybo A, Mudrak O (2003) Etymological dictionary of the Altaic languages. Brill, Leiden

Suzuki H (2022) Geolinguistics in the eastern Tibetosphere: an introduction. Geolinguistic Society of Japan, Tokyo. https://doi.org/10.5281/zenodo.5989176

Tao L, Yuan H, Zhu K et al (2023) Ancient genomes reveal millet farming-related demic diffusion from the Yellow River into southwest China. Curr Biol 33(22):4995–5002.E7. https://doi.org/10.1016/j.cub.2023.09.055

Tournadre N, Suzuki H (2023) The Tibetic languages: an introduction to the family of languages derived from Old Tibetan. LACITO Publications, Villejuif. https://doi.org/10.5281/zenodo.10026628

Wang W, Mauleon R, Hu Z et al (2018) Genomic variation in 3, 010 diverse accessions of Asian cultivated rice. Nature 557:43–49. https://doi.org/10.1038/s41586-018-0063-9

Watanabe T (2010) Mugi wo arawasu ko-kanji [Old Chinese characters which denoted wheat]. In: Sato Y et al (eds) Mugi no shizenshi [Natural history of wheat]. Hokkaido Daigaku Shuppankai, Sapporo, pp 27–34. [in Japanese]

Whitman J (2011) Northeast Asian linguistic ecology and the advent of rice agriculture in Korea and Japan. Rice 4:149–158. https://doi.org/10.1007/s12284-011-9080-0

Zhang S (2019) Three ancient words for bear. Sino-Plat Pap 294:1–21. https://sino-platonic.org/complete/spp294_words_for_bear.pdf

Zhengzhang S (1990) Shanggu rusheng yunwei de qingzhuo wenti [Problem on voicedness of the entering tone endings in Old Chinese]. Yuyan Yanjiu 1990(1):67–74. [in Chinese]

Zohary D, Weiss E, Hopf M (2012) Cereals. In: Domestication of plants in the Old World: the origin and spread of domesticated plants in Southwest Asia, Europe, and the Mediterranean Basin, 4th edn. Oxford University Press, Oxford, pp 20–74. https://doi.org/10.1093/acprof:osobl/9780199549061.003.0003

Index

N. Osada et al. (eds.), *Phylogeographic History of Plants and Animals Coexisting with Humans in Asia*, Evolutionary Studies,
https://doi.org/10.1007/978-981-97-6887-5

SPRINGER NATURE

GPSR Compliance

The European Union's (EU) General Product Safety Regulation (GPSR) is a set of rules that requires consumer products to be safe and our obligations to ensure this.

If you have any concerns about our products, you can contact us on ProductSafety@springernature.com

In case Publisher is established outside the EU, the EU authorized representative is:

Springer Nature Customer Service Center GmbH
Europaplatz 3
69115 Heidelberg, Germany

The manufacturer's authorised representative in the EU is Springer Nature Customer Service Centre GmbH, Europaplatz 3, 69115 Heidelberg, Germany. If you have any concerns regarding our products, please contact ProductSafety@springernature.com

Printed and bound by CPI Group (UK) Ltd, Croydon, CR0 4YY

30/09/2026

02225240-0001